Differential Calculus Statistics and Probability

A Well Simplified Math Book for high Schools and Colleges

Kingsley Augustine

Printed by Amazon KDP

Table of Contents

PREFACE

Differential Calculus, Statistics and Probability is a book that contains topics under differential calculus, statistics and probability. This book serves as a useful companion for students in high schools and higher institutions of learning. It is a valuable textbook for students who want to write entrance test or examination into colleges and universities. This book consists of step-by-step explanation of topics presented in a way that is easy for students to understand. It contains very many worked examples and many self-assessment exercises to ensure that students get a mastery of each topic covered. The answers to the exercises are provided at the end of the book.

What makes this book a unique mathematical asset, is its detailed step by step approach in explaining the topics covered in these branch of mathematics. Instead of solving questions by going straight to the point, leaving you confused and frustrated, this textbook teaches you in simple English, explaining each step taken at a time. Thus, allowing anyone, regardless of their experience in differential calculus, statistics and probability, to understand each topic with ease, and hence make mathematics more interesting.

I give all thanks and Glory to God Almighty, for giving me the grace to write this book. I also wish to express my deep appreciation to my wife Mrs. Mercy Augustine for her patience, understanding and encouragement when I was writing this book. I also thank my children, Dora, Merit and Elvis for their moral support.

Kingsley Augustine.

CHAPTER 1
LIMIT OF A FUNCTION

The concept of limits is very important in differential calculus. A function has a limiting value when its variable approaches a certain value. For example, if the limiting value of f(x) as x approaches 4 is 16, it is written as:

$$\lim_{x \to 4} f(x) = 16$$

In evaluating limit, some problems are simply solved by just putting in the values of the variables, while some problems are solved by applying certain rules.

Some important limits

1. $\lim_{x \to 0} \dfrac{\sin x}{x} = 1$ and $\lim_{x \to 0} \dfrac{\tan x}{x} = 1$ (Or $\lim_{x \to 0} \dfrac{\sin ax}{ax} = 1$ and $\lim_{x \to 0} \dfrac{\tan ax}{ax} = 1$)

2. $\lim_{x \to 0} \dfrac{1 - \cos x}{x} = 0$ and $\lim_{x \to 0} \dfrac{1 - \cos a x}{x} = 0$

3. $\lim_{x \to \infty} \left(1 + \dfrac{1}{x}\right)^x = e$

4. $\lim_{x \to 0} \dfrac{a^x - 1}{a^x} = 1$

5. $\lim_{x \to 0} \dfrac{e^x - 1}{x} = 1$

6. $\lim_{x \to 0} \dfrac{1 + x}{x} = 1$

7. $\lim_{x \to a} \dfrac{x^n - a^n}{x - a} = na^{n-1}$

8. $\lim_{x \to 0} \dfrac{a^x - 1}{x} = \ln a$

9. $\lim_{x \to \infty} \dfrac{\ln x}{x} = 0$

10. $\lim_{x \to \infty} x^{\frac{1}{x}} = 0$

11. $\lim_{x \to 0} (1 + x)^{\frac{1}{x}} = e$

12. $\lim_{x \to 0} (1 + \sin x)^{\frac{1}{x}} = e$

13. $\lim_{x \to a} c = c$

Examples

1. Evaluate $\lim\limits_{x\to 0} 3x^3 - 5x^2 + 2x + 6$

<u>Solution</u>

$$\lim\limits_{x\to 0} 3x^3 - 5x^2 + 2x + 6$$

Simply substitute zero for x in the given expression. This gives:

$$= 3(0^3) - 5(0^2) + 2(0) + 6$$
$$= 0 - 0 + 0 + 6$$
$$= 6$$

2. Evaluate $\lim\limits_{x\to 0} \dfrac{x^2 + 3x - 11}{4x^2 - 5x - 5}$

<u>Solution</u>

$$\lim\limits_{x\to 0} \dfrac{x^2 + 3x - 11}{4x^2 - 5x - 5}$$

Substituting 0 for x into the expression gives:

$$= \dfrac{0^2 + 3(0) - 11}{4(0)^2 - 5(0) - 5}$$
$$= \dfrac{0 + 0 - 11}{0 - 0 - 5}$$
$$= \dfrac{-11}{-5}$$
$$= \dfrac{11}{5}$$
$$= 2\dfrac{1}{5}$$

3. Evaluate $\lim\limits_{x\to 4} \dfrac{x^2 - 2x - 8}{x - 4}$

<u>Solution</u>

A close look at the expression shows that if 4 is substituted for x, it will give $\dfrac{0}{0}$. This has no value as it is indeterminate. Therefore, in order to solve a limit such as this, we have to factorize the denominator and then simplify the expression. This is done as follows:

$$\lim\limits_{x\to 4} \dfrac{x^2 - 2x - 8}{x - 4} = \lim\limits_{x\to 4} \dfrac{(x+2)(x-4)}{x - 4}$$

Cancelling out $(x - 4)$ gives:

$$\lim\limits_{x\to 4} (x + 2)$$

We now substitute 4 for x to obtain our final answer as follows:

$$\lim\limits_{x\to 4} (x + 2) = 4 + 2$$
$$= 6$$

Therefore, $\lim\limits_{x\to 4} \dfrac{x^2 + x - 8}{x - 4} = 6$

4. Evaluate $\lim_{x \to 2} \dfrac{x^2 - 4}{x - 2}$

<u>Solution</u>

This is similar to example 3 above since substituting 2 for x in the expression will $\dfrac{0}{0}$ which has no value. Hence we factorize the numerator and simplify as follows:

$$\lim_{x \to 2} \frac{x^2 - 4}{x - 2} = \lim_{x \to 2} \frac{(x-2)(x+2)}{x - 2} \quad \text{[Note that } x^2 - 4 = x^2 - 2^2 \text{ and recall that a}^2 - b^2 = (a + b)(a - b)\text{]}$$

Cancelling out $(x - 2)$ gives:

$$\lim_{x \to 2} (x + 2)$$

We now substitute 2 for x to obtain our final answer as follows:

$$\lim_{x \to 2} (x + 2) = 2 + 2$$
$$= 4$$

Therefore, $\lim_{x \to 2} \dfrac{x^2 - 4}{x - 2} = 4$

5. Evaluate $\lim_{x \to 0} (2x - 5)(x + 2)(3x - 1)$

<u>Solution</u>

$$\lim_{x \to 0} (2x - 5)(x + 2)(3x - 1)$$

Substituting 0 for x in the expression gives:

$$\lim_{x \to 0} (2x - 5)(x + 2)(3x - 1) = [2(0) - 5][(0) + 2][3(0) - 1]$$
$$= (-5)(2)(-1)$$
$$= 10$$

6. Evaluate $\lim_{x \to \infty} \dfrac{x^3 + 2x^2 + 5x + 3}{x^3 + x^2 + 3x + 7}$

<u>Solution</u>

In this case of x approaching infinity, we evaluate it by first dividing each term in the numerator and denominator by the variable having the highest exponent (power). Hence we divide each term by x^3. This is done as follows:

$$\lim_{x \to \infty} \frac{x^3 + 2x^2 + 5x + 3}{x^3 + x^2 + 3x + 7} = \lim_{x \to \infty} \frac{\frac{x^3}{x^3} + \frac{2x^2}{x^3} + \frac{5x}{x^3} + \frac{3}{x^3}}{\frac{x^3}{x^3} + \frac{x^2}{x^3} + \frac{3x}{x^3} + \frac{7}{x^3}}$$

$$= \lim_{x \to \infty} \frac{1 + \frac{2}{x} + \frac{5}{x^2} + \frac{3}{x^3}}{1 + \frac{1}{x} + \frac{3}{x^2} + \frac{7}{x^3}}$$

$$= \frac{1 + \frac{2}{\infty} + \frac{5}{\infty^2} + \frac{3}{\infty^3}}{1 + \frac{1}{\infty} + \frac{3}{\infty^2} + \frac{7}{\infty^3}}$$

$$= \frac{1+0+0+0}{1+0+0+0} \qquad \text{(Note that a number divided by } \infty \text{ gives 0)}$$

$$= \frac{1}{1}$$

$$= 1$$

7. Evaluate $\lim\limits_{x\to\infty} \dfrac{2x^4 - 3x^2 + 8x - 1}{x^4 - 5x + 3}$

Solution

The variable in its highest exponent (power) is x^4. Hence we divide each term by x^4 as follows:

$$\lim_{x\to\infty} \frac{2x^4 - 3x^2 + 8x - 1}{x^4 - 5x + 3} = \lim_{x\to\infty} \frac{\frac{2x^4}{x^4} - \frac{3x^2}{x^4} + \frac{8x}{x^4} - \frac{1}{x^4}}{\frac{x^4}{x^4} - \frac{5x}{x^4} + \frac{3}{x^4}}$$

$$= \lim_{x\to\infty} \frac{2 - \frac{3}{x^2} + \frac{8}{x^3} - \frac{1}{x^4}}{1 - \frac{5}{x^3} + \frac{3}{x^4}}$$

$$= \frac{2 - \frac{3}{\infty} + \frac{8}{\infty} - \frac{1}{\infty}}{1 - \frac{5}{\infty} + \frac{3}{\infty}}$$

$$= \frac{2 - 0 + 0 - 0}{1 - 0 + 0} \qquad \text{(Note that a number divided by } \infty \text{ gives 0)}$$

$$= \frac{2}{1}$$

$$= 2$$

8. Evaluate $\lim\limits_{x\to 4} \dfrac{2x^2 - 7x - 4}{x^2 - 3x - 4}$

Solution

If 4 is substituted for x in the function above, it will give $\dfrac{0}{0}$ which is an indeterminate value.

Hence we factorize the expression and simplify as follows:

$$\lim_{x\to 4} \frac{2x^2 - 7x - 4}{x^2 - 3x - 4} = \lim_{x\to 4} \frac{(2x+1)(x-4)}{(x+1)(x-4)}$$

Cancelling out $(x - 4)$ gives:

$$\lim_{x\to 4} \frac{2x + 1}{x + 1}$$

We now substitute 4 for x to obtain our final answer as follows:

$$\lim_{x\to 4} \frac{2x + 1}{x + 1} = \frac{2(4) + 1}{(4) + 1}$$

$$= \frac{8+1}{4+1}$$

$$= \frac{9}{5}$$

$$= 1\frac{4}{5}$$

Therefore, $\lim\limits_{x \to 4} \dfrac{2x^2 - 7x - 4}{x^2 - 3x - 4} = 1\frac{4}{5}$

9. Evaluate $\lim\limits_{x \to 3} \dfrac{x^2 - 9}{x^2 + 6x + 9}$

Solution

Substituting 3 for x in the function can give us the final answer as follows:

$$\lim\limits_{x \to 3} \frac{x^2 - 9}{x^2 + 6x + 9} = \frac{3^2 - 9}{3^2 + 6(3) + 9}$$

$$= \frac{9 - 9}{9 + 18 + 9}$$

$$= \frac{0}{36}$$

$$= 0$$

Hence, $\lim\limits_{x \to 3} \dfrac{x^2 - 9}{x^2 + 6x + 9} = 0$

Note that even if this expression had been factorized and simplified, the final answer would still give zero.

10. Determine the limiting value of $\dfrac{x + 2}{x^2 + 6x + 8}$ as x tends to -2 .

Solution

The question can also be written as $\lim\limits_{x \to -2} \dfrac{x + 2}{x^2 + 6x + 8}$

If -2 is substituted for x in the function above, it will give $\dfrac{0}{0}$. Hence we factorize the expression and simplify as follows:

$$\lim\limits_{x \to -2} \frac{x + 2}{x^2 + 6x + 8} = \lim\limits_{x \to -2} \frac{x + 2}{(x + 2)(x + 4)}$$

Cancelling out $(x + 2)$ gives:

$$\lim\limits_{x \to -2} \frac{1}{x + 4}$$

We now substitute -2 for x as follows:

$$\lim\limits_{x \to -2} \frac{1}{x + 4} = \frac{1}{-2 + 4}$$

$$= \frac{1}{2}$$

11. Find the limit of $\dfrac{x^3 - 125}{x - 5}$ as $x \to 5$

<u>Solution</u>

The question can also be written as $\displaystyle\lim_{x \to 5} \dfrac{x^3 - 125}{x - 5}$

We factorize the expression and simplify as follows:

$$\lim_{x \to 5} \frac{x^3 - 125}{x - 5} = \lim_{x \to 5} \frac{x^3 - 5^3}{x - 5}$$

$$= \lim_{x \to 5} \frac{(x - 5)(x^2 + 5x + 5^2)}{x - 5} \qquad \text{[Recall the identity } a^3 - b^3 = (a - b)(a^2 + ab + b^2)]$$

Cancelling out $(x - 5)$ gives:

$$\lim_{x \to 5} (x^2 + 5x + 5^2)$$

We now substitute 5 for x as follows:

$$\lim_{x \to 5} (x^2 + 5x + 5^2) = (5^2 + 5(5) + 25)$$

$$= 25 + 25 + 25$$

$$= 75$$

12. Find the limit of $\dfrac{x^3 + 64}{x + 4}$ as $x \to -4$

<u>Solution</u>

The question can also be written as $\displaystyle\lim_{x \to -4} \dfrac{x^3 + 4^3}{x + 4}$

We factorize the expression and simplify as follows:

$$\lim_{x \to -4} \frac{x^3 + 64}{x + 4} = \lim_{x \to -4} \frac{x^3 + 4^3}{x + 4}$$

$$= \lim_{x \to -4} \frac{(x + 4)(x^2 - 4x + 4^2)}{x + 4} \qquad \text{[Recall the identity } a^3 + b^3 = (a + b)(a^2 - ab + b^2)]$$

Cancelling out $(x + 4)$ gives:

$$\lim_{x \to -4} (x^2 - 4x + 4^2)$$

We now substitute -4 for x as follows:

$$\lim_{x \to -4} (x^2 - 4x + 4^2) = (-4^2 - 4(-4) + 16)$$

$$= 16 + 16 + 16$$

$$= 48$$

13. Evaluate $\displaystyle\lim_{x \to 9} \dfrac{3 - \sqrt{x}}{9 - x}$

<u>Solution</u>

If 9 is substituted for x in the expression above, it will give $\dfrac{0}{0}$ which is not the desired answer.

Hence a way to go around this problem is to multiply the top and bottom by the conjugate of the surd at the top. The conjugate of $3 - \sqrt{x}$ is $3 + \sqrt{x}$ (only a difference in their middle signs).

Hence we multiply top and bottom by $3 + \sqrt{x}$ and simplify as follows:

$$\lim_{x \to 9} \frac{3 - \sqrt{x}}{9 - x} = \lim_{x \to 9} \frac{(3 - \sqrt{x})(3 + \sqrt{x})}{(9 - x)(3 + \sqrt{x})}$$

$$= \lim_{x \to 9} \frac{3^2 - (\sqrt{x})^2}{(9 - x)(3 + \sqrt{x})} \qquad \text{(Note that the top was simplified by using } (a - b)(a + b) = a^2 - b^2)$$

$$= \lim_{x \to 9} \frac{9 - x}{(9 - x)(3 + \sqrt{x})}$$

$$= \lim_{x \to 9} \frac{1}{(3 + \sqrt{x})} \qquad \text{(Since } 9 - x \text{ cancels out)}$$

We now substitute 9 for x as follows:

$$\lim_{x \to 9} \frac{1}{(3 + \sqrt{x})} = \frac{1}{(3 + \sqrt{9})}$$

$$= \frac{1}{3 + 3}$$

$$= \frac{1}{6}$$

14. Evaluate $\lim_{x \to 1} \dfrac{x^2 - 1}{x^5 - 1}$

Solution

If 2 is substituted for x in the expression above, it will give $\dfrac{0}{0}$ which is not a good answer. Hence a way to go around this problem is to divide the top and bottom by $x - 1$ (i.e. the difference between the variable and 1). This is done as follows:

$$\lim_{x \to 1} \frac{x^2 - 1}{x^5 - 1} = \lim_{x \to 1} \frac{\dfrac{x^2 - 1}{x - 1}}{\dfrac{x^5 - 1}{x - 1}} \qquad \text{(Note that this has not changed the original fraction)}$$

Note that $\lim_{x \to a} \dfrac{x^n - a^n}{x - a} = na^{n-1}$. With this rule we now simplify the expression above as follows:

$$\lim_{x \to 1} \frac{\dfrac{x^2 - 1}{x - 1}}{\dfrac{x^5 - 1}{x - 1}} = \frac{2 \times 1^{(2 - 1)}}{5 \times 1^{(5 - 1)}}$$

$$= \frac{2 \times 1^1}{5 \times 1^5}$$

$$= \frac{2}{5}$$

Continuity of a function

A function is said to be continuous if the three conditions below are satisfied.

1. f(a) exists

2. $\displaystyle\lim_{x \to a} f(x)$ exists

3. $\displaystyle\lim_{x \to a} f(x) = f(a)$ (i.e. if the values of conditions 1 and 2 above are equal)

Examples

1. Determine if the function $f(x) = 2x^2 - 8x + 5$ is continuous at the point $x = -1$

Solution

When $x = -1$, the $f(-1)$ is obtained as follows:

$$f(x) = 2x^2 - 8x + 5$$
$$f(-1) = 2(-1)^2 - 8(-1) + 5$$
$$= 2 + 8 + 5$$
$$= 15$$

Since $f(-1) = 15$, it means that $f(-1)$ exists.

The next step is to find the value of $\displaystyle\lim_{x \to -1} f(x)$ as follows:

$$\lim_{x \to -1} 2x^2 - 8x + 5 = 2(-1)^2 - 8(-1) + 5$$
$$= 2 + 8 + 5$$
$$= 15$$

Hence $f(-1) = \displaystyle\lim_{x \to -1} f(x) = 15$

Therefore, the function is continuous.

2. Determine if the function $f(x) = \dfrac{2x + 1}{x^2 - 3x + 7}$ is continuous at $x = 2$

Solution

When $x = 2$, the $f(2)$ is obtained as follows:

$$f(x) = \frac{2x + 1}{x^2 - 3x + 7}$$
$$f(2) = \frac{2(2) + 1}{2^2 - 3(2) + 7}$$
$$= \frac{4 + 1}{4 - 6 + 7}$$
$$= \frac{5}{5}$$
$$= 1$$

Since $f(2) = 1$, it means that $f(2)$ exists.

The next step is to find the value of $\displaystyle\lim_{x \to 2} f(x)$ as follows:

$$\lim_{x \to 2} \frac{2x + 1}{x^2 - 3x + 7} = \frac{2(2) + 1}{2^2 - 3(2) + 7}$$
$$= \frac{4 + 1}{4 - 6 + 7}$$
$$= \frac{5}{5}$$

$$= 1$$

Hence $f(2) = \underset{x \to 2}{lim} f(x) = 1$

Therefore, the function is continuous.

3. Determine if the function $f(x) = \dfrac{2x^2 - 8}{x + 4}$ is continuous at $x = -4$.

When $x = -4$, the f(-4) is obtained as follows:

$$f(x) = \frac{2x^2 - 8}{x + 4}$$

$$f(-4) = \frac{2(-4^2) - 8}{-4 + 4}$$

$$= \frac{2(16) - 8}{0}$$

$$= \frac{24}{0} \quad \text{(Undefined)}$$

Hence f(-4) is undefined, and does not exist. This shows that the function is not continuous. It is discontinuous as $x = -4$.

4. Determine if the function $f(x) = \dfrac{x^2 - 9}{x - 3}$ is continuous at $x = 3$.

When $x = 3$, the f(3) is obtained as follows:

$$f(x) = \frac{x^2 - 9}{x - 3}$$

$$f(3) = \frac{3^2 - 9}{3 - 3}$$

$$= \frac{0}{0} \quad \text{(This has no value)}$$

Hence f(3) does not exist.

Let us determine the value of $\underset{x \to 3}{lim} f(x)$ as follows:

$$\underset{x \to 3}{lim} \frac{x^2 - 9}{x - 3} = \underset{x \to 3}{lim} \frac{(x - 3)(x + 3)}{x - 3}$$

$$= \underset{x \to 3}{lim} (x + 3) \quad \text{(Since } x - 3 \text{ cancels out)}$$

$$= 3 + 3 \quad \text{(When 3 is substituted for } x\text{)}$$

$$= 6$$

Hence, f(3) has no value while $\underset{x \to 3}{lim} f(x) = 6$

This shows that $f(3) \neq \underset{x \to 3}{lim} f(x)$

Therefore the function discontinuous.

5. Test for the continuity of the function: $f(x) = \begin{cases} -5x + 7 & \text{if } x > 1 \\ x^2 + 2 & \text{if } x \leq 1 \end{cases}$

Solution

$$f(x) = \begin{cases} -5x + 7 & \text{if } x > 1 \\ x^2 + 2 & \text{if } x \leq 1 \end{cases}$$

15

When $x = 1$, we use the expression $x^2 + 2$ to evaluate f(1). Note that $x \leq 1$ means $x < 1$ or $x = 1$.

Hence, $f(x) = x^2 + 2$ (for $x = 1$)

$$f(1) = 1^2 + 2$$
$$= 3$$

Since f(1) = 3, it means that f(1) exist.

Let us now determine the limiting value of f(x).

If x approaches 1 from the left, which is $x \leq 1$ (left hand elbow) we write it as $\lim_{x \to 1^-} f(x)$, and we pick the expression given by:

$$f(x) = x^2 + 2$$

Hence, $\lim_{x \to 1} f(x) = 1^2 + 3$

$$= 3$$

If x approaches 1 from the right, which is $x > 1$ (right hand elbow), we write it as $\lim_{x \to 1^+} f(x)$, and we pick the expression given by:

$$f(x) = -5x + 7$$

Hence, $\lim_{x \to 1} f(x) = -5(1) + 7$

$$= 2$$

Hence, $\lim_{x \to 1^-} f(x) = 3$ while $\lim_{x \to 1^+} f(x) = 2$. They are not equal. This shows that $\lim_{x \to 1} f(x)$ does not exist. Therefore, the function is not continuous.

6. Determine the continuity of the function: $f(x) = \begin{cases} x^2 & \text{if } x \geq -2 \\ x+6 & \text{if } x < -2 \end{cases}$

Solution

$$f(x) = \begin{cases} x^2 & \text{if } x \geq -2 \\ x+6 & \text{if } x < -2 \end{cases}$$

A direct way of determining the continuity of functions such as these is to find the limits of the two expressions in the function as x tends to the values given in the question.

Therefore, when x tends to -2 in the function $f(x) = x^2$, we have:

$$f(x) = x^2$$

Hence, $\lim_{x \to -2} f(x) = (-2)^2$

$$= 4$$

Also, when x tends to -2 in the function $f(x) = x + 6$, we have:

$$f(x) = x + 6$$

Hence, $\lim_{x \to -2} f(x) = -2 + 6$

$$= 4$$

Hence, $\lim_{x \to -2} x^2 = \lim_{x \to -2} x + 6 = 4$. Their limits give the same value. This shows that $\lim_{x \to -2} f(x)$ exist. Therefore, the function is continuous.

7. Determine if the function below is continuous or not.

$$f(x) = \begin{cases} \dfrac{x^2 - 16}{x + 4} & \text{if } x \geq -4 \\ \dfrac{x + 4}{2x - 3} & \text{if } x < -4 \end{cases}$$

Solution

As x tends to -4 in the function $f(x) = \dfrac{x^2 - 16}{x + 4}$ we simplify as follows:

$$f(x) = \frac{(x + 4)(x - 4)}{x + 4}$$

Hence, $\lim\limits_{x \to -4} f(x) = (x - 4)$ (Note that $(x + 4)$ cancels out)

$$= -4 - 4$$
$$= -8$$

Also, when x tends to -4 in the function $f(x) = \dfrac{x + 4}{2x - 3}$, we have:

$$\lim\limits_{x \to -4} f(x) = \frac{-4 + 4}{2(-4) - 3}$$
$$= \frac{-4 + 4}{-8 - 3}$$
$$= \frac{0}{-11}$$
$$= 0$$

Their limits give different values. This shows that $\lim\limits_{x \to -4} f(x)$ does not exist. Therefore, the function is not continuous.

8. If $f(x) = \begin{cases} \dfrac{x^2 - 4}{x - 2} \\ 4 \;\; \text{if } x = 2 \end{cases}$ test for the continuity of the function at $x = 2$.

Solution

$$f(x) = \begin{cases} \dfrac{x^2 - 4}{x - 2} \\ 4 \;\; \text{if } x = 2 \end{cases}$$

Let us find the limits of the two expressions in the function as x tends 2.

From the first expression, as x tends to 2 we have:

$$f(x) = \frac{x^2 - 4}{x - 2}$$

Hence, $\lim\limits_{x \to 2} f(x) = \dfrac{(x - 2)(x + 2)}{x - 2}$ (When the numerator is factorized)

$\lim\limits_{x \to 2} f(x) = (x + 2)$ ($x - 2$ has cancelled out)

$\lim\limits_{x \to 2} f(x) = 2 + 2$

$$= 4$$

Also, when x tends to 2 in the function $f(x) = 4$, we have:

$$f(x) = 4$$

17

Hence, $\lim_{x \to 2} 4 = 4$ (Since $\lim_{x \to a} c = c$)

The question also tells us that when $x = 2$, $f(x) = 4$

Hence, $\lim_{x \to 2} 4 = \lim_{x \to 2} \dfrac{x^2 - 4}{x - 2} = 4$. Their limits give the same value of 4. This shows that f(x) exist,

and that $f(2) = \lim_{x \to 2} f(x)$. Therefore, the function is continuous.

Examples on Limits of Trigonometric Functions

1. Evaluate $\lim_{x \to 0} \dfrac{\sin 2x}{5x}$

Solution

$$\lim_{x \to 0} \dfrac{\sin 2x}{5x}$$

Substituting zero directly into the expression above will give $\dfrac{0}{0}$. Hence we have to apply the

appropriate rule of limit. Let us first make an adjustment to the expression as follows.

Multiply the expression by $\dfrac{2x}{5x}$ and change the denominator to $2x$ just like the numerator, as

follows:

$$\lim_{x \to 0} \dfrac{\sin 2x}{5x} = \lim_{x \to 0} \left(\dfrac{\sin 2x}{2x} \times \dfrac{2x}{5x} \right)$$

Note that when $2x$ cancels out, the original expression remains the same.

Recall that $\lim_{x \to 0} \dfrac{\sin ax}{ax} = 1$

Hence, $\lim_{x \to 0} \dfrac{\sin 2x}{2x} = 1$

Substituting 1 for $\lim_{x \to 0} \dfrac{\sin 2x}{2x}$ in the simplification given above to gives:

$$\lim_{x \to 0} \left(\dfrac{\sin 2x}{2x} \times \dfrac{2x}{5x} \right) = \lim_{x \to 0} \dfrac{\sin 2x}{2x} \times \lim_{x \to 0} \dfrac{2x}{5x}$$

$$= 1 \times \lim_{x \to 0} \dfrac{2}{5} \quad (x \text{ cancels out})$$

$$= 1 \times \dfrac{2}{5} \quad (\text{Since } \lim_{x \to a} c = c)$$

$$= \dfrac{2}{5}$$

2. 1. Evaluate $\lim_{x \to 0} \dfrac{\sin 5x}{7x}$

Solution

$$\lim_{x \to 0} \dfrac{\sin 5x}{7x}$$

Multiply the expression by $\dfrac{5x}{7x}$ and change the denominator to $5x$ just like the numerator. This

gives:

$$\lim_{x \to 0} \frac{\sin 5x}{7x} = \lim_{x \to 0} \left(\frac{\sin 5x}{5x} \times \frac{5x}{7x} \right)$$

Hence, $\lim_{x \to 0} \dfrac{\sin 5x}{5x} = 1$

Substituting 1 for $\lim_{x \to 0} \dfrac{\sin 5x}{5x}$ in the simplification given above to gives:

$$\lim_{x \to 0} \left(\frac{\sin 5x}{5x} \times \frac{5x}{7x} \right) = \lim_{x \to 0} \frac{\sin 5x}{5x} \times \lim_{x \to 0} \frac{5x}{7x}$$

$$= 1 \times \lim_{x \to 0} \frac{5}{7} \quad (x \text{ cancels out})$$

$$= 1 \times \frac{5}{7} \quad (\text{Since } \lim_{x \to a} c = c)$$

$$= \frac{5}{7}$$

The two examples above show that $\lim_{x \to 0} \dfrac{\sin a\, x}{x} = a$ or $\lim_{x \to 0} \dfrac{\sin a\, x}{bx} = \dfrac{a}{b}$. This rule also applies to tangent as $\lim_{x \to 0} \dfrac{\tan a\, x}{x} = a$ or $\lim_{x \to 0} \dfrac{\tan a\, x}{bx} = \dfrac{a}{b}$.

3. Find the value of $\lim_{x \to \frac{\pi}{2}} \dfrac{\cos 2x}{\sin 3x}$

Solution

$$\lim_{x \to \frac{\pi}{2}} \frac{\cos 2x}{\sin 3x}$$

We can simply substitute in the value of x and obtain our answer.

$$\lim_{x \to \frac{\pi}{2}} \frac{\cos 2x}{\sin 3x} = \frac{\cos 2(\frac{\pi}{2})}{\sin 3(\frac{\pi}{2})}$$

$$= \frac{\cos \pi}{\sin \frac{3\pi}{2}}$$

$$= \frac{\cos 180}{\sin 270} \quad (\text{Note that } \pi \text{ radians} = 180^\circ)$$

$$= \frac{-1}{-1}$$

$$= 1$$

This problem can also be solved by using the angles directly in radians, but I prefer to work in degrees. Note that cos π in radians = −1, and $\sin \dfrac{3\pi}{2}$ in radians = −1 as obtained above.

4. Evaluate $\lim_{x \to 0} \dfrac{\cos x}{\sin x - 3}$

Solution

$$\lim_{x \to 0} \frac{\cos x}{\sin x - 3} = \frac{\cos 0}{\sin 0 - 3}$$

$$= \frac{1}{0-3}$$

$$= -\frac{1}{3}$$

5. Evaluate $\lim\limits_{x \to 0} \dfrac{\sin 3x}{\sin 2x}$

Solution

Multiply by $\dfrac{x}{\sin 2x}$ and change the denominator to x as follows.

$$\lim\limits_{x \to 0} \frac{\sin 3x}{\sin 2x} = \lim\limits_{x \to 0} \left(\frac{\sin 3x}{x} \text{ X } \frac{x}{\sin 2x} \right) \quad \text{(When } x \text{ cancels out it gives the original expression)}$$

$$= \lim\limits_{x \to 0} \left(\frac{\sin 3x}{x} \text{ X } \left(\frac{\sin 2x}{x}\right)^{-1} \right) \quad \text{(Note that the inverse of } \frac{x}{\sin 2x} \text{ has been taken)}$$

$$= \lim\limits_{x \to 0} \frac{\sin 3x}{x} \text{ X } \left(\lim\limits_{x \to 0} \frac{\sin 2x}{x} \right)^{-1}$$

$$= 3 \text{ x } 2^{-1} \quad \text{(Note that } \lim\limits_{x \to 0} \frac{\sin 3x}{x} = 3 \text{ and } \lim\limits_{x \to 0} \frac{\sin 2x}{x} = 2 \text{)}$$

$$= 3 \text{ x } \frac{1}{2}$$

$$= \frac{3}{2}$$

This example shows that: $\lim\limits_{x \to 0} \dfrac{\sin a\,x}{\sin b\,x} = \dfrac{a}{b}$

6. Evaluate $\lim\limits_{x \to 0} \dfrac{\sin 2x}{\sin 5x}$

Solution

$$\lim\limits_{x \to 0} \frac{\sin 2x}{\sin 5x}$$

From the rule established in example 5 above, we can see that the solution to this problem is $\dfrac{2}{5}$.

Hence, $\lim\limits_{x \to 0} \dfrac{\sin 2x}{\sin 5x} = \dfrac{2}{5}$ $\quad$ (Since $\lim\limits_{x \to 0} \dfrac{\sin a\,x}{\sin b\,x} = \dfrac{a}{b}$)

Exercise 1

1. Evaluate $\lim\limits_{x \to 0} 2x^3 - 4x^2 + x + 9$

2. Evaluate $\lim\limits_{x \to 0} \dfrac{5x^2 + x - 8}{x^2 - 2x + 5}$

3. Evaluate $\lim\limits_{x \to 5} \dfrac{x^2 - x - 20}{x - 5}$

4. Evaluate $\lim\limits_{x \to 7} \dfrac{x^2 - 49}{x - 7}$

5. Evaluate $\lim\limits_{x \to 0} (5x - 1)(3x + 7)(3x + 2)$

6. Evaluate $\lim\limits_{x \to \infty} \dfrac{2x^3 + x^2 - 9x - 5}{5x^3 + 2x^2 - 7x + 2}$

7. Evaluate $\lim\limits_{x \to \infty} \dfrac{x^4 - 5x^2 - 2x - 6}{3x^4 + 2x + 3}$

8. Evaluate $\lim\limits_{x \to 4} \dfrac{x^2 - 25}{x^2 + 2x - 15}$

9. Evaluate $\lim\limits_{x \to 8} \dfrac{2x^2 - 17x + 8}{8 - x}$

10. Evaluate $\lim\limits_{m \to 0} \dfrac{(5+m)^2 - 25}{m}$

11. Determine the limiting value of $\dfrac{\sqrt{x} - 2}{x - 4}$ as x tends to 4 .

12. Find the limit of $\dfrac{x^3 - 27}{x - 3}$ as $x \to 3$

13. Find the limit of $\dfrac{x^3 + 8}{x + 2}$ as $x \to -2$

14. Evaluate $\lim\limits_{x \to 25} \dfrac{5 - \sqrt{x}}{25 - x}$

15. Evaluate $\lim\limits_{x \to 1} \dfrac{x^3 - 1}{x^7 - 1}$

16. Determine if the function $f(x) = x^2 + 3x - 1$ is continuous at the point $x = 2$

17. Determine if the function $f(x) = \dfrac{5x - 2}{3x^2 - x - 2}$ is continuous at $x = -1$

18. Determine if the function $f(x) = \dfrac{2x^2 - 18}{x - 3}$ is continuous at $x = 3$.

19. Determine if the function $f(x) = \dfrac{x^3 + 64}{x + 4}$ is continuous at $x = -4$.

20. Test for the continuity of the function: $f(x) = \begin{cases} -x+5 & \text{if } x > 2 \\ x^3 + 10 & \text{if } x \le 2 \end{cases}$

21. Determine the continuity of the function: $f(x) = \begin{cases} 3x^2 & \text{if } x \ge -3 \\ x+30 & \text{if } x < -3 \end{cases}$

22. Determine if the function below is continuous or not.

$f(x) = \begin{cases} \dfrac{x^2 - 25}{x + 5} & \text{if } x \ge -5 \\ \dfrac{x + 5}{2x + 10} & \text{if } x < -5 \end{cases}$

23. If $f(x) = \begin{cases} \dfrac{x^3 - 27}{x - 3} \\ 1 \ \text{if } x = 3 \end{cases}$ test for the continuity of the function at $x = 3$.

24. If $f(x) = \begin{cases} 2 - 3x & x < 1 \\ x^3 + 4 & x \ge 1 \end{cases}$ evaluate the following limits, if they exist.

(a) $\lim\limits_{x \to -2} f(x)$

(b) $\lim\limits_{x \to 1} f(x)$

25. Evaluate $\lim\limits_{x \to 2} (5 + |x - 2|)$ if it exists.

26. Evaluate $\lim\limits_{x \to 0} \dfrac{\sin 7x}{2x}$

27. Evaluate $\lim\limits_{x \to 0} \dfrac{\sin x}{3x}$

28. Find the value of $\lim\limits_{x \to \pi} \dfrac{\cos 2x}{\sin \frac{x}{2}}$

29. Evaluate $\lim\limits_{x \to 0} \dfrac{\cos 5x}{\sin 3x - 2}$

30. Evaluate $\lim\limits_{x \to 0} \dfrac{\sin 9x}{\sin 4x}$

CHAPTER 2
DIFFERENTIATION FROM FIRST PRINCIPLE

If $y = f(x)$, then the gradient function of y or $f(x)$ is given by:

$$g(x) = \frac{f(x + \Delta x) - f(x)}{\Delta x}$$

where Δx is the increment in x.

Applying limits has shown that as Δx tends to zero,

$$f'(x) = \lim_{\Delta x \to 0} \frac{f(x + \Delta x) - f(x)}{\Delta x}$$

$$\frac{dy}{dx} = \lim_{\Delta x \to 0} \frac{f(x + \Delta x) - f(x)}{\Delta x}$$

$\dfrac{dy}{dx}$ reads dee y dee x.

Note that $f'(x)$ and $\dfrac{dy}{dx}$ can be interpreted as instantaneous rate of change of y with respect to x.

The technique of finding the derivative of a function by considering the limiting value is called differentiation from first principle.

Examples

1. Find the derivative of x from first principle.

Solution

Let, $y = x$

Increasing x by Δx will give an increase in y by Δy. This gives:

$$y + \Delta y = x + \Delta x$$
$$\Delta y = x + \Delta x - y$$
$$\Delta y = x + \Delta x - x \quad \text{(substitute } x \text{ for y since y} = x \text{ as given above from the question)}$$
$$\Delta y = \Delta x \quad (x \text{ cancels out})$$

Dividing both sides by Δx in order to get the derivative gives:

$$\frac{\Delta y}{\Delta x} = \frac{\Delta x}{\Delta x}$$
$$\frac{\Delta y}{\Delta x} = 1$$

Taking limits as Δx tends to zero gives:

$$\lim_{\Delta x \to 0} \frac{\Delta y}{\Delta x} = 1 \quad \text{[Recall that } \lim_{x \to 0} c = c, \text{ hence } \lim_{\Delta x \to 0} f(\Delta x) = 1 \text{ gives 1, since } \frac{\Delta y}{\Delta x} \text{ is regarded as } f(\Delta x)]$$

We now write $\lim\limits_{\Delta x \to 0} \dfrac{\Delta y}{\Delta x}$ as $\dfrac{dy}{dx}$. This gives:

$$\frac{dy}{dx} = 1$$

2. Find from first principle, the derivative of $2x^3$.

<u>Solution</u>

Let, $y = 2x^3$

Increasing x by Δx will give an increase in y by Δy. This gives:

$$y + \Delta y = 2(x + \Delta x)^3$$

$$y + \Delta y = 2[x^3 + 3x^2\Delta x + 3x(\Delta x)^2 + (\Delta x)^3]$$

$$= 2x^3 + 6x^2\Delta x + 6x(\Delta x)^2 + 2(\Delta x)^3$$

$$\Delta y = 2x^3 + 6x^2\Delta x + 6x(\Delta x)^2 + 2(\Delta x)^3 - y$$

$$= 2x^3 + 6x^2\Delta x + 6x(\Delta x)^2 + 2(\Delta x)^3 - 2x^3 \quad \text{(Since } y = 2x^3\text{)}$$

$$\Delta y = 6x^2\Delta x + 6x(\Delta x)^2 + 2(\Delta x)^3 \quad \text{(Since } 2x^3 \text{ cancels out)}$$

Dividing both sides by Δx in order to get the derivative gives:

$$\frac{\Delta y}{\Delta x} = \frac{6x^2\Delta x}{\Delta x} + \frac{6x(\Delta x)^2}{\Delta x} + \frac{2(\Delta x)^3}{\Delta x}$$

$$\frac{\Delta y}{\Delta x} = 6x^2 + 6x\Delta x + 2(\Delta x)^2$$

Taking limits as Δx tends to zero gives:

$$\lim_{\Delta x \to 0} \frac{\Delta y}{\Delta x} = 6x^2 + 6x(0) + 2(0)^2 \quad \text{(Note that zero replaces } \Delta x\text{)}$$

$$\lim_{\Delta x \to 0} \frac{\Delta y}{\Delta x} = 6x^2$$

We now write $\lim_{\Delta x \to 0} \frac{\Delta y}{\Delta x}$ as $\frac{dy}{dx}$. This gives:

$$\frac{dy}{dx} = 6x^2$$

3. Find the derivative of $f(x) = \dfrac{1}{x^2}$ from first principle.

<u>Solution</u>

$$f(x) = \Delta x$$

Increasing x by Δx gives:

$$f(x + \Delta x) = \frac{1}{(x + \Delta x)^2}$$

$$f(x + \Delta x) = \frac{1}{x^2 + 2x\Delta x + (\Delta x)^2}$$

Subtract $f(x)$ from both sides of the equation

$$f(x + \Delta x) - f(x) = \frac{1}{x^2 + 2x\Delta x + (\Delta x)^2} - f(x)$$

$$f(x + \Delta x) - f(x) = \frac{1}{x^2 + 2x\Delta x + (\Delta x)^2} - \frac{1}{x^2} \quad \text{(Note that } f(x) = \frac{1}{x^2}\text{)}$$

$$= \frac{x^2 - (x^2 + 2x\Delta x + (\Delta x)^2)}{x^2(x^2 + 2x\Delta x + (\Delta x)^2)}$$

$$= \frac{x^2 - x^2 - 2x\Delta x - (\Delta x)^2}{x^2(x^2 + 2x\Delta x + (\Delta x)^2)}$$

$$= \frac{-2x\Delta x - (\Delta x)^2}{x^2(x^2 + 2x\Delta x + (\Delta x)^2)}$$

$$f(x + \Delta x) - f(x) = \frac{\Delta x(-2x - \Delta x)}{x^2(x^2 + 2x\Delta x + (\Delta x)^2)}$$

Dividing both sides by Δx gives:

$$\frac{f(x + \Delta x) - f(x)}{\Delta x} = \frac{-2x - \Delta x}{x^2(x^2 + 2x\Delta x + (\Delta x)^2)} \qquad (\Delta x \text{ has cancelled out from the right hand side})$$

Taking limits as Δx tends to zero gives:

$$\lim_{\Delta x \to 0} \frac{f(x + \Delta x) - f(x)}{\Delta x} = \frac{-2x - 0}{x^2(x^2 + 2x(0) + (0)^2)}$$

Replacing $\displaystyle\lim_{\Delta x \to 0} \frac{f(x + \Delta x) - f(x)}{\Delta x}$ with f'(x) gives the derivative of f(x) as:

$$f'(x) = \frac{-2x}{x^2(x^2)}$$

$$f'(x) = \frac{-2}{x^3}$$

4. If $f(x) = 3x^2$

(a) write down and simplify the expression $\dfrac{f(x + h) - f(x)}{h}$ (h $\neq$ 0)

(b) find $\displaystyle\lim_{\Delta x \to 0} \frac{f(x + h) - f(x)}{h}$

<u>Solution</u>

(a) Let, $f(x) = 3x^2$

Increasing x by h gives:

$$f(x + h) = 3(x + h)^2$$
$$= 3[x^2 + 2xh + h^2]$$
$$= 3x^2 + 6xh + 3h^2$$

Subtracting f(x) from both sides gives:

$$f(x + h) - f(x) = 3x^2 + 6xh + 3h^2 - 3x^2 \qquad \text{(Note that } f(x) = 3x^2 \text{ from the question)}$$
$$= 6xh + 3h^2$$

Dividing both side by h gives:

$$\frac{f(x + h) - f(x)}{h} = \frac{6xh}{h} + \frac{3h^2}{h}$$

$$\frac{f(x + h) - f(x)}{h} = 6x + 3h$$

(b) $\dfrac{f(x + h) - f(x)}{h} = 6x + 3h$

Taking limits as h tends to zero gives:

$$\lim_{h \to 0} \frac{f(x + h) - f(x)}{h} = 6x + 3(0)$$

$$\lim_{h \to 0} \frac{f(x + h) - f(x)}{h} = 6x$$

25

5. If $f(x) = 4x^3 - 5$,

(a) evaluate $\dfrac{f(x+h) - f(x)}{h}$, where $h \neq 0$

(b) From your result in (5a) above, find the derivatives of $f(x)$ with respect to x.

Solution

(a) Let, $f(x) = 4x^3 - 5$

Increasing x by h gives:

$$f(x + h) = 4(x + h)^3 - 5$$
$$= 4[x^3 + 3x^2h + 3xh^2 + h^3] - 5$$
$$= 4x^3 + 12x^2h + 12xh^2 + 4h^3 - 5$$

Subtracting $f(x)$ from both sides gives:

$$f(x + h) - f(x) = 4x^3 + 12x^2h + 12xh^2 + 4h^3 - 5 - (4x^3 - 5) \quad (f(x) = 4x^3 - 5 \text{ from the question})$$
$$= 4x^3 + 12x^2h + 12xh^2 + 4h^3 - 5 - 4x^3 + 5$$
$$= 12x^2h + 12xh^2 + 4h^3$$

Dividing both side by h gives:

$$\frac{f(x+h) - f(x)}{h} = 12x^2 + 12xh + 4h^2$$

(b) $\dfrac{f(x+h) - f(x)}{h} = 12x^2 + 12xh + 4h^2$

Taking limits as h tends to zero gives:

$$\lim_{h \to 0} \frac{f(x+h) - f(x)}{h} = 12x^2 + 12x(0)\, 4(0)^2$$

$$\lim_{h \to 0} \frac{f(x+h) - f(x)}{h} = 12x^2$$

6. If $f(x) = \dfrac{x^2 + 1}{2x}$

(a) write down and simplify the expression $\dfrac{f(x + \Delta x) - f(\Delta x)}{\Delta x}$, where $\Delta x \neq 0$

(b) Find the derivatives of $y = \dfrac{x^2 + 1}{2x}$

Solution

$$f(x) = \frac{x^2 + 1}{2x}$$

$$f(x + \Delta x) = \frac{(x + \Delta x)^2 + 1}{2(x + \Delta x)}$$

$$f(x + \Delta x) = \frac{x^2 + 2x\Delta x + (\Delta x)^2 + 1}{2x + 2\Delta x}$$

Subtract $f(x)$ from both sides of the equation

$$f(x + \Delta x) - f(x) = \frac{x^2 + 2x\Delta x + (\Delta x)^2 + 1}{2x + 2\Delta x} - \frac{x^2 + 1}{2x}$$

26

$$f(x + \Delta x) - f(x) = \frac{2x[x^2 + 2x\Delta x + (\Delta x)^2 + 1] - (x^2 + 1)(2x + 2\Delta x)}{2x(2x + 2\Delta x)}$$

$$= \frac{2x^3 + 4x^2\Delta x + 2x(\Delta x)^2 + 2x - (2x^3 + 2x^2\Delta x + 2x + 2\Delta x)}{2x(2x + 2\Delta x)}$$

$$= \frac{2x^3 + 4x^2\Delta x + 2x(\Delta x)^2 + 2x - 2x^3 - 2x^2\Delta x - 2x - 2\Delta x}{2x(2x + 2\Delta x)}$$

$$= \frac{2x^2\Delta x + 2x(\Delta x)^2 - 2\Delta x}{2x(2x + 2\Delta x)}$$

$$f(x + \Delta x) - f(x) = \frac{\Delta x(2x^2 + 2x\Delta x - 2)}{2x(2x + 2\Delta x)} \qquad \text{(After factorizing the numerator)}$$

Dividing both sides by Δx gives:

$$\frac{f(x + \Delta x) - f(x)}{\Delta x} = \frac{2x^2 + 2x\Delta x - 2}{2x(2x + 2\Delta x)}$$

(b) Taking limits as Δx tends to zero gives:

$$\lim_{\Delta x \to 0} \frac{f(x + \Delta x) - f(x)}{\Delta x} = \frac{2x^2 + 2x(0) - 2}{2x(2x + 2(0))}$$

$$= \frac{2x^2 - 2}{2x(2x)}$$

$$= \frac{2(x^2 - 1)}{4x^2}$$

Replacing $\lim_{\Delta x \to 0} \frac{f(x + \Delta x) - f(x)}{\Delta x}$ with $f'(x)$ gives the derivative of $f(x)$ as:

$$f'(x) = \frac{x^2 - 1}{2x^2}$$

Separating into fractions gives:

$$f'(x) = \frac{x^2}{2x^2} - \frac{1}{2x^2}$$

$$f'(x) = \frac{1}{2} - \frac{1}{2x^2}$$

7. Find from first principle the derivatives with respect to x of $y = 5x^3 - x + 7$

Solution

$$y = 5x^3 - x + 7$$

$$y + \Delta y = 5(x + \Delta x)^3 - (x + \Delta x) + 7$$

$$y + \Delta y = 5[x^3 + 3x^2\Delta x + 3x(\Delta x)^2 + (\Delta x)^3] - x - \Delta x + 7$$

$$= 5x^3 + 15x^2\Delta x + 15x(\Delta x)^2 + 5(\Delta x)^3 - x - \Delta x + 7$$

$$\Delta y = 5x^3 + 15x^2\Delta x + 15x(\Delta x)^2 + 5(\Delta x)^3 - x - \Delta x + 7 - y$$

$$= 5x^3 + 15x^2\Delta x + 15x(\Delta x)^2 + 5(\Delta x)^3 - x - \Delta x + 7 - (5x^3 - x + 7) \quad \text{(Since } y = 5x^3 - x + 7\text{)}$$

$$\Delta y = 5x^3 + 15x^2\Delta x + 15x(\Delta x)^2 + 5(\Delta x)^3 - x - \Delta x + 7 - 5x^3 + x - 7$$

$$= 15x^2\Delta x + 15x(\Delta x)^2 + 5(\Delta x)^3 - \Delta x$$

Dividing both sides by Δx gives:

$$\frac{\Delta y}{\Delta x} = \frac{15x^2\Delta x}{\Delta x} + \frac{15x(\Delta x)^2}{\Delta x} + \frac{5(\Delta x)^3}{\Delta x} - \frac{\Delta x}{\Delta x}$$

$$\frac{\Delta y}{\Delta x} = 15x^2 + 15x\Delta x + 5(\Delta x)^2 - 1$$

Taking limits as Δx tends to zero gives:

$$\lim_{\Delta x \to 0} \frac{\Delta y}{\Delta x} = 15x^2 + 15x(0) + 5(0)^2 - 1$$

$$\lim_{\Delta x \to 0} \frac{\Delta y}{\Delta x} = 15x^2 - 1$$

We now replace $\lim_{\Delta x \to 0} \frac{\Delta y}{\Delta x}$ with $\frac{dy}{dx}$. This gives:

$$\frac{dy}{dx} = 15x^2 - 1$$

8. Differentiate $3x - \dfrac{1}{2x^2}$ from first principle.

<u>Solution</u>

$$y = 3x - \frac{1}{x^2}$$

$$y + \Delta y = 3(x + \Delta x) - \frac{1}{2(x + \Delta x)^2}$$

$$= 3x + 3\Delta x - \frac{1}{2[x^2 + 2x\Delta x + (\Delta x)^2]}$$

Subtract y from both sides of the equation

$$\Delta y = 3x + 3\Delta x - \frac{1}{2x^2 + 4x\Delta x + 2(\Delta x)^2} - y$$

$$= 3x + 3\Delta x - \frac{1}{2x^2 + 4x\Delta x + 2(\Delta x)^2} - \left(3x - \frac{1}{2x^2}\right)$$

$$= 3x + 3\Delta x - \frac{1}{2x^2 + 4x\Delta x + 2(\Delta x)^2} - 3x + \frac{1}{2x^2}$$

$$= 3\Delta x - \frac{1}{2x^2 + 4x\Delta x + 2(\Delta x)^2} + \frac{1}{2x^2}$$

Combining them into one fraction gives:

$$\Delta y = \frac{3\Delta x[2x^2(2x^2 + 4x\Delta x + 2(\Delta x)^2] - 2x^2 + 2x^2 + 4x\Delta x + 2(\Delta x)^2}{2x^2[2x^2 + 4x\Delta x + 2(\Delta x)^2]}$$

$$= \frac{3\Delta x[4x^4 + 8x^3\Delta x + 4x^2(\Delta x)^2] + 4x\Delta x + 2(\Delta x)^2}{2x^2[2x^2 + 4x\Delta x + 2(\Delta x)^2]} \quad (-2x^2 + 2x^2 \text{ has cancelled out})$$

$$= \frac{12x^4\Delta x + 24x^3(\Delta x)^2 + 12x^2(\Delta x)^3 + 4x\Delta x + 2(\Delta x)^2}{2x^2[2x^2 + 4x\Delta x + 2(\Delta x)^2]}$$

Factorizing the numerator gives:

$$= \frac{\Delta x[12x^4 + 24x^3\Delta x + 12x^2(\Delta x)^2 + 4x + 2\Delta x]}{2x^2[2x^2 + 4x\Delta x + 2(\Delta x)^2]}$$

Dividing both sides by Δx gives:

$$\frac{\Delta y}{\Delta x} = \frac{12x^4 + 24x^3\Delta x + 12x^2(\Delta x)^2 + 4x + 2\Delta x}{2x^2[2x^2 + 4x\Delta x + 2(\Delta x)^2]}$$ (Δx that is outside the bracket cancels out)

Taking limits as Δx tends to zero gives:

$$\lim_{\Delta x \to 0} \frac{\Delta y}{\Delta x} = \frac{12x^4 + 24x^3(0) + 12x^2(0)^2 + 4x + 2(0)}{2x^2[2x^2 + 4x(0) + 2(0)^2]}$$

$$= \frac{12x^4 + 4x}{2x^2[2x^2]}$$

$$\lim_{\Delta x \to 0} \frac{\Delta y}{\Delta x} = \frac{12x^4 + 4x}{4x^4}$$

$$= \frac{12x^4}{4x^4} + \frac{4x}{4x^4}$$

$$= 3 + \frac{1}{x^3}$$

Replacing $\lim_{\Delta x \to 0} \frac{\Delta y}{\Delta x}$ with $\frac{dy}{dx}$ gives the derivative as:

$$\frac{dy}{dx} = 3 + \frac{1}{x^3}$$

Exercise 2

1. Find the derivative of $2x$ from first principle.

2. Find from first principle, the derivative of x^2.

3. Find the derivative of $f(x) = \frac{1}{x^3}$ from first principle.

4. If $f(x) = 5x^2$

(a) write down and simplify the expression $\frac{f(x+h) - f(x)}{h}$ ($h \neq 0$)

(b) find $\lim_{h \to 0} \frac{f(x+h) - f(x)}{h}$

5. If $f(x) = 9x^3$,

(a) evaluate $\frac{f(x+h) - f(x)}{h}$, where $h \neq 0$

(b) From your result above, find the derivatives of $f(x)$ with respect to x.

6. If $f(x) = \frac{x^3 - 2}{x}$

(a) write down and simplify the expression $\frac{f(x+\Delta x) - f(\Delta x)}{\Delta x}$, where $\Delta x \neq 0$

(b) Find the derivatives of $y = \frac{x^3 - 2}{x}$

7. Find from first principle the derivatives with respect to x of $y = 3x^2 - 10x$

8. Differentiate $x + \dfrac{3}{x}$ from first principle.

9. From first principle, find the derivatives with respect to x of $y = 5x - 3x^2$

10. Differentiate $2x + \dfrac{x}{5}$ from first principle.

CHAPTER 3
GENERAL RULE OF DIFFERENTIATION AND COMPOSITE FUNCTIONS

The general rule for the derivative/differentiation of a function is as given below.

If $y = x^n$

then, $\dfrac{dy}{dx} = nx^{n-1}$

The rule for differentiating a composite function (or function of a function) is given as follows:

If $y = (2x + 5)^4$

then we write, $u = 2x + 5$

and express y as:

$y = u^4$

Therefore, $\dfrac{dy}{dx} = \dfrac{dy}{du} \times \dfrac{du}{dx}$

This rule is called the chain rule.

Examples

1. Find the derivatives of the following:

(a) $y = 2x^7$

(b) $y = \dfrac{3}{4}x^8$

(c) $y = 3\sqrt{x}$

(d) $y = \dfrac{10}{\sqrt[5]{x^3}}$

(e) $y = \dfrac{1}{x^{\frac{1}{4}}}$

Solution

(a) $y = 2x^7$

Solutions

(a) $y = 2x^7$

$\dfrac{dy}{dx} = 7 \times 2x^{7-1}$

This is done by multiplying the exponent (power) by the term and subtracting 1 from the exponent (power). Hence, the answer is:

$\dfrac{dy}{dx} = 14x^6$

(b) $y = \dfrac{3}{4}x^8$

Multiply the term by the exponent (power) (i.e. 8) and subtract 1 from the exponent. This gives:

$\dfrac{dy}{dx} = 8 \times \dfrac{3}{4}x^{8-1}$

$$= \frac{24}{4} x^7$$

$$\frac{dy}{dx} = 6x^7$$

(c) $y = 3\sqrt{x}$

Since $\sqrt{x} = x^{\frac{1}{2}}$, we can rewrite the expression as:

$$y = 3x^{\frac{1}{2}}$$

$$\frac{dy}{dx} = \frac{1}{2} \times 3x^{\frac{1}{2} - 1}$$

$$= \frac{3}{2} x^{-\frac{1}{2}}$$

$$= \frac{3}{2} \times \frac{1}{x^{\frac{1}{2}}} \qquad \text{(Note that } x^{-\frac{1}{2}} = \frac{1}{x^{\frac{1}{2}}} \text{ from indices)}$$

$$\frac{dy}{dx} = \frac{3}{2x^{\frac{1}{2}}}$$

$$\frac{dy}{dx} = \frac{3}{2\sqrt{x}} \qquad \text{(Since } x^{\frac{1}{2}} = \sqrt{x} \text{)}$$

(d) $y = \frac{10}{\sqrt[5]{x^3}}$

Expressing the root in fractional form gives:

$$y = \frac{10}{(x^3)^{\frac{1}{5}}}$$

$$y = \frac{10}{x^{\frac{3}{5}}} \qquad \text{(The two exponents (powers) 3 and } \frac{1}{5} \text{ have been multiplied)}$$

Taking the denominator to the numerator changes the sign of its exponent as follows:

$$y = 10x^{-\frac{3}{5}}$$

Hence, $\dfrac{dy}{dx} = -\dfrac{3}{5} \times 10x^{-\frac{3}{5} - 1}$

$$= -\frac{30}{5} x^{-\frac{8}{5}}$$

$$= -6 x^{-\frac{8}{5}}$$

$$= -6 \times \frac{1}{x^{\frac{8}{5}}} \qquad \text{(Note that the inverse of a term changes the sign of its exponent)}$$

$$\frac{dy}{dx} = \frac{-6}{x^{\frac{8}{5}}}$$

Or, $\dfrac{dy}{dx} = \dfrac{-6}{\sqrt[5]{x^8}}$

(e) $y = \dfrac{1}{x^{\frac{1}{4}}}$

This can be expressed as:

$$y = x^{-\frac{1}{4}}$$

$$\dfrac{dy}{dx} = -\dfrac{1}{4}x^{-\frac{1}{4}-1}$$

$$= -\dfrac{1}{4}x^{-\frac{5}{4}}$$

$$= -\dfrac{1}{4} \times \dfrac{1}{x^{\frac{5}{4}}}$$

$$\dfrac{dy}{dx} = -\dfrac{1}{4x^{\frac{5}{4}}} \qquad \text{(Note that the inverse of a term changes the sign of its exponent)}$$

Or, $\dfrac{dy}{dx} = \dfrac{1}{4\sqrt[4]{x^5}}$

2. Find the derivative of each of the following:

(a) $5x^3 - 7x^2 - 3x + 8$

(b) $\dfrac{3}{5}x^5 + 2x^3 - x$

(c) $\dfrac{2x^4 - 5x^3 - 4x^2 + 3}{x^2}$

(d) $\sqrt{x} + \dfrac{1}{\sqrt{x}}$

Solutions

(a) Let the expression be $y = 5x^3 - 7x^2 - 3x + 8$

Hence, $\dfrac{dy}{dx} = \dfrac{d(5x^3)}{dx} - \dfrac{d(7x^2)}{dx} - \dfrac{d(3x)}{dx} + \dfrac{d(8)}{dx}$

This means that each part should be differentiated separately.

Hence, $\dfrac{dy}{dx} = (3 \times 5x^{3-1}) - (2 \times 7x^{2-1}) - (1 \times 3x^{1-1}) + 0$

$\dfrac{dy}{dx} = 15x^2 - 14x - 3$ (Note that $x^0 = 1$)

Note that the derivative of a constant is zero as shown by the derivative of 8

(b) $y = \frac{3}{5}x^5 + 2x^3 - x$

$\quad\quad \frac{dy}{dx} = (5 \times \frac{3}{5}x^{5-1}) + (3 \times 2x^{3-1}) - (1 \times x^{1-1})$

$\quad\quad\quad = 3x^4 + 6x^2 - 1$

(c) $y = \dfrac{2x^4 - 5x^3 - 4x^2 + 3}{x^2}$

Dividing each term in the numerator by the denominator in order to separate the expression into it different fractions gives:

$\quad y = \dfrac{2x^4}{x^2} - \dfrac{5x^3}{x^2} - \dfrac{4x^2}{x^2} + \dfrac{3}{x^2}$

$\quad\quad = 2x^2 - 5x - 4 + \dfrac{3}{x^2}$

$\quad y = 2x^2 - 5x - 4 + 3x^{-2}$

$\quad \dfrac{dy}{dx} = (2 \times 2x) - 5 - 0 + (-2 \times 3x^{-2-1})$

$\quad\quad = 4x - 5 - 6x^{-3}$

$\quad \dfrac{dy}{dx} = 4x - 5 - \dfrac{6}{x^2}$

(d) $y = \sqrt{x} + \dfrac{1}{\sqrt{x}}$

$\quad\quad = x^{\frac{1}{2}} + \dfrac{1}{x^{\frac{1}{2}}}$

$\quad y = x^{\frac{1}{2}} + x^{-\frac{1}{2}}$

$\quad \dfrac{dy}{dx} = \dfrac{1}{2}x^{-\frac{1}{2}} - \dfrac{1}{2}x^{-\frac{3}{2}}$ $\quad$ (Note that $-\dfrac{1}{2} - 1 = -\dfrac{3}{2}$)

$\quad\quad = \dfrac{1}{2} \times \dfrac{1}{x^{\frac{1}{2}}} - \dfrac{1}{2} \times \dfrac{1}{x^{\frac{3}{2}}}$

$\quad\quad = \dfrac{1}{2x^{\frac{1}{2}}} - \dfrac{1}{2x^{\frac{3}{2}}}$

$\quad \dfrac{dy}{dx} = \dfrac{1}{2\sqrt{x}} - \dfrac{1}{2\sqrt{x^3}}$

3. If $y = (5x - 2)^3$, find $\dfrac{dy}{dx}$

Solution

$\quad\quad y = (5x - 2)^3$ $\quad$ (This is a composite function)

34

Let us take u = 5x – 2

If 5x – 2 is replaced with U, then the question (i.e. y = $(5x – 2)^3$) becomes:

$$y = u^3$$

Hence, $\frac{dy}{du} = 3u^2$

Since u = 5x – 2

then $\frac{du}{dx} = 5$

Therefore, $\frac{dy}{dx} = \frac{dy}{du} \times \frac{du}{dx}$ (Chain rule)

$$= 3u^2 \times 5$$

$$= 15u^2$$

Now substitute 5x – 2 for u to obtain $\frac{dy}{dx}$ as follows:

$$\frac{dy}{dx} = 15(5x – 2)^2$$

4. If y = $\frac{1}{(5x^2 – 1)^4}$ find $\frac{dy}{dx}$

Solution

$$y = \frac{1}{(5x^2 – 1)^4}$$ (This is a composite function)

It can also be represented as follows:

$$y = (5x^2 – 1)^{-4}$$ (Its inverse changes the sign of its exponent)

Now, let us take u = $5x^2$ – 1

Hence, y = u^{-4} (When $5x^2$ – 1 is replaced with u in the original question)

Therefore, $\frac{dy}{du} = -4u^{-5}$

Since u = $5x^2$ – 1

then, $\frac{du}{dx} = 10x$

Therefore, $\frac{dy}{dx} = \frac{dy}{du} \times \frac{du}{dx}$ (Chain rule)

$$= -4u^{-5} \times 10x$$

$$\frac{dy}{dx} = -40xu^{-5}$$

Now substitute $5x^2$ – 1 for u. This gives:

$$\frac{dy}{dx} = -40x(5x^2 – 1)^{-5}$$

Or, $\frac{dy}{dx} = \frac{-40x}{(5x^2 – 1)^5}$ (Note the change in the sign of the exponent as it becomes

denominator)

5. If $y = (2x^3 + 7x)^{\frac{1}{2}}$ find $\dfrac{dy}{dx}$

<u>Solution</u>

$$y = (2x^3 + 7x)^{\frac{1}{2}}$$

Let $u = 2x^3 + 7x$

Hence, $y = u^{\frac{1}{2}}$

$$\frac{dy}{du} = \frac{1}{2}u^{-\frac{1}{2}}$$

Also, $u = 2x^3 + 7x$

$$\frac{du}{dx} = 6x^2 + 7$$

Therefore, $\dfrac{dy}{dx} = \dfrac{dy}{du} \times \dfrac{du}{dx}$

$$= \frac{1}{2}u^{-\frac{1}{2}} \times 6x^2 + 7$$

$$= \frac{6x^2+7}{2}u^{-\frac{1}{2}}$$

$$= \frac{6x^2+7}{2} \times \frac{1}{u^{1/2}}$$

$$\frac{dy}{dx} = \frac{6x^2+7}{2u^{1/2}}$$

Now, replace u with $2x^3 + 7x$. This gives:

$$\frac{dy}{dx} = \frac{6x^2+7}{2(2x^3+7x)^{1/2}}$$

Or, $\dfrac{dy}{dx} = \dfrac{6x^2+7}{2\sqrt{2x^3+7x}}$ (Note that $(2x^3 + 7x)^{\frac{1}{2}} = \sqrt{2x^3 + 7x}$)

6. Find the derivative of $3x^2 - x + 9)^4$

<u>Solution</u>

$$y = (3x^2 - x + 9)^4$$

Let $u = 3x^2 - x + 9$

Hence, $y = u^4$

$$\frac{dy}{du} = 4u^3$$

$$\frac{du}{dx} = 6x - 1$$

Therefore, $\dfrac{dy}{dx} = \dfrac{dy}{du} \times \dfrac{du}{dx}$

$$= 4u^3 \times 6x - 1$$

$$= 4(6x - 1)u^3$$

$$= (24x - 4)u^3$$

$$\frac{dy}{dx} = (24x - 4)(3x^2 - x + 9)^3 \quad \text{(When u is replaced with } 3x^2 - x + 9\text{)}$$

7. Find the derivative of $\left(x - \dfrac{5}{x}\right)^4$

Solution

$$y = \left(x - \frac{5}{x}\right)^4$$

Let $u = x - \dfrac{5}{x}$

Therefore, $y = u^4$

$$\frac{dy}{du} = 4u^3$$

$$\frac{du}{dx} = \frac{d(x)}{dx} - \frac{d(5x^{-1})}{dx} \quad \text{(Note that } = \frac{5}{x} = 5x^{-1}\text{)}$$

$$= 1 - (-1)5x^{-2}$$

$$= 1 + 5x^{-2}$$

$$\frac{du}{dx} = 1 + \frac{5}{x^2}$$

Therefore, $\dfrac{dy}{dx} = \dfrac{dy}{du} \times \dfrac{du}{dx}$

$$= 4u^3 \times \left(1 + \frac{5}{x^2}\right)$$

$$= \left(4 + \frac{20}{x^2}\right)u^3$$

$$\frac{dy}{dx} = \left(4 + \frac{20}{x^2}\right)\left(x - \frac{5}{x}\right)^3 \quad \text{(When u is replaced with } \left(x - \frac{5}{x}\right)\text{)}$$

8. Differentiate with respect to x: $\dfrac{1}{2x^5 - 3x + 1}$

Solution

Let $y = \dfrac{1}{2x^5 - 3x + 1}$

Or, $y = (2x^5 - 3x + 1)^{-1} \quad$ (Recall from indices that $\dfrac{1}{a} = a^{-1}$)

Let us take $u = 2x^5 - 3x + 1$

Therefore, $y = u^{-1}$

$$\frac{dy}{du} = -1u^{-2}$$

$$\frac{du}{dx} = 10x^4 - 3$$

Therefore, $\dfrac{dy}{dx} = \dfrac{dy}{du} \times \dfrac{du}{dx}$

$$= -1u^{-2} \times 10x^4 - 3$$

$$= -1(10x^4 - 3)u^{-2}$$

$$= (-10x^4 + 3)u^{-2}$$

$$= \frac{3 - 10x^4}{u^2}$$

$$\frac{dy}{dx} = \frac{3 - 10x^4}{(2x^5 - 3x + 1)^2}$$

9. Differentiate with respect to x: $\sqrt{7 - 5x^3}$

<u>Solution</u>

Let $y = \sqrt{7 - 5x^3}$

Or, $y = (7 - 5x^3)^{\frac{1}{2}}$ (Recall from indices that $\sqrt{a} = a^{\frac{1}{2}}$)

Let $u = 7 - 5x^3$

Hence, $y = u^{\frac{1}{2}}$

$$\frac{dy}{du} = \frac{1}{2} u^{-\frac{1}{2}}$$

$$\frac{du}{dx} = -15x^2$$

Therefore, $\dfrac{dy}{dx} = \dfrac{dy}{du} \times \dfrac{du}{dx}$

$$= \frac{1}{2} u^{-\frac{1}{2}} \times (-15x^2)$$

$$= -\frac{15}{2} x^2 (u^{-\frac{1}{2}})$$

$$= -\frac{15x^2}{2u^{\frac{1}{2}}}$$

$$= -\frac{15x^2}{2\sqrt{u}}$$

$$\frac{dy}{dx} = -\frac{15x^2}{2\sqrt{7 - 5x^3}}$$

10. Find $\dfrac{dy}{dx}$ if $y = \dfrac{1}{\sqrt{2x^3 - 5}}$

<u>Solution</u>

$$y = \frac{1}{\sqrt{2x^3 - 5}}$$

Let $u = 2x^3 - 5$

hence, $y = \dfrac{1}{\sqrt{u}}$

$$y = \frac{1}{u^{\frac{1}{2}}}$$

$$y = u^{-\frac{1}{2}}$$

$$\frac{dy}{du} = -\frac{1}{2} u^{-\frac{3}{2}} \quad \text{(Note that } -\frac{1}{2} - 1 = -\frac{3}{2}\text{)}$$

38

$$\frac{du}{dx} = 6x^2$$

Therefore, $\dfrac{dy}{dx} = \dfrac{dy}{du} \times \dfrac{du}{dx}$

$$= -\frac{1}{2}u^{-\frac{3}{2}} \times (6x^2)$$

$$= -\frac{6}{2}x^2\,(u^{-\frac{3}{2}})$$

$$= -3x^2(u^{-\frac{3}{2}})$$

$$= -\frac{3x^2}{u^{\frac{3}{2}}}$$

$$= -\frac{3x^2}{\sqrt{u^3}}$$

$$\frac{dy}{dx} = -\frac{3x^2}{\sqrt{(2x^3-5)^3}}$$

Exercise 3

1. Find the derivatives of the following:

(a) $y = 8x^5$

(b) $y = \dfrac{2}{5}x^5$

(c) $y = \sqrt[3]{x}$

(d) $y = 7\sqrt[7]{x}$

(e) $y = \dfrac{1}{\sqrt[8]{x^5}}$

(f) $y = \dfrac{2}{x^{\frac{5}{2}}}$

2. Find the derivative of each of the following:

(a) $2x^5 - 3x^4 - 4x^3 + 5x^2 - 6x + 7$

(b) $x^7 + 2x^4 - \dfrac{3}{x}$

(c) $\dfrac{3x^9 - x^7 - 5x^4 + 2x^2 - 1}{x^3}$

(d) $5(\sqrt[4]{x}) + \dfrac{5}{\sqrt[3]{2x}}$

3. If $y = (2x - 5)^4$, find $\dfrac{dy}{dx}$

4. If $y = \dfrac{3}{(x^3 - 7)^2}$ find $\dfrac{dy}{dx}$

5. If $y = (2x^3 + 7x)^{\frac{1}{2}}$ find $\dfrac{dy}{dx}$

6. Find the derivative of $(7x^3 - x^2 + 3)^5$

7. Find the derivative of $\left(3x - \dfrac{2}{3x}\right)^3$

8. Differentiate with respect to x: $\quad -\dfrac{9}{3x^2 - x - 10}$

9. Differentiate with respect to x: $\sqrt{1 - 2x^4}$

10. Find $\dfrac{dy}{dx}$ if $y = \dfrac{1}{\sqrt[3]{5x^3 - 1}}$

11. Find the derivative of $(x^5 - 3)^9$

12. Find the derivative of $\left(x - \dfrac{1}{5x}\right)^2$

13. Differentiate with respect to x: $\dfrac{2}{x^3 - x - \frac{1}{x}}$

14. Differentiate with respect to x: $\sqrt{5x - x^2}$

15. Find $\dfrac{dy}{dx}$ if $y = \dfrac{1}{\sqrt[5]{x^3 + 2}}$

CHAPTER 4
PRODUCT RULE OF DERIVATIVE

If $y = uv$ where u and v are functions of x, then:

$$\frac{dy}{dx} = u\frac{dv}{dx} + v\frac{du}{dx}$$

This is called the product rule of differentiation.

Similarly, If $y = uvw$ where u, v and w are functions of x, then:

$$\frac{dy}{dx} = \frac{du}{dx}vw + \frac{dv}{dx}uw + \frac{dw}{dx}uv$$

Examples

1. If $f(x) = (x - 3)(x + 4)$, find $f'(x)$

Solution

$f(x) = (x - 3)(x + 4)$

This is a product of two functions of x.

Let $u = x - 3$

and $v = x + 4$

Hence, $\dfrac{du}{dx} = 1$

$\dfrac{dv}{dx} = 1$

Therefore, the derivative of $f(x)$ is:

$$f'(x) = u\frac{dv}{dx} + v\frac{du}{dx}$$

$$= (x - 3) \times 1 + (x + 4) \times 1$$

$$= x - 3 + x + 4$$

$$f'(x) = 2x + 1$$

Note that another way to differentiate the function in example 1 above is to expand the bracket and differentiate directly.

2. Find the derivative of $y = (4x^2 + 1)(x^2 - 3)$

Solution

$y = (4x^2 + 1)(x^2 - 3)$

Let $u = 4x^2 + 1$

and $v = (x^2 - 3)$

Hence, $\dfrac{du}{dx} = 8x$

$\dfrac{dv}{dx} = 2x$

$$\frac{dy}{dx} = u\frac{dv}{dx} + v\frac{du}{dx}$$

$$= (4x^2 + 1)2x + (x^2 - 3)8x$$

$$= 8x^3 + 2x + 8x^3 - 24x$$

$$\frac{dy}{dx} = 16x^3 - 22x$$

3. If $y = x^2(1 + 2x)^{\frac{1}{2}}$, find $\frac{dy}{dx}$

Solution

$$y = x^2(1 + 2x)^{\frac{1}{2}}$$

$$u = x^2$$

and $v = (1 + 2x)^{\frac{1}{2}}$

Hence, $\frac{du}{dx} = 2x$

$$\frac{dv}{dx} = \frac{1}{2} \times 2 \times (1 + 2x)^{\frac{1}{2} - 1} \qquad \text{(Use of chain rule)}$$

$$= (1 + 2x)^{-\frac{1}{2}}$$

$$\frac{dv}{dx} = \frac{1}{(1+2x)^{\frac{1}{2}}}$$

Hence, $\frac{dy}{dx} = u\frac{dv}{dx} + v\frac{du}{dx}$

$$= x^2 \frac{1}{(1+2x)^{\frac{1}{2}}} + (1 + 2x)^{\frac{1}{2}} \times 2x$$

$$= \frac{x^2}{(1+2x)^{\frac{1}{2}}} + 2x(1 + 2x)^{\frac{1}{2}}$$

let us now simplify by using $(1 + 2x)^{\frac{1}{2}}$ as the LCM as follows:

$$= \frac{x^2 + 2x\ (1+2x)^{\frac{1}{2}}(1+2x)^{\frac{1}{2}}}{(1+2x)^{\frac{1}{2}}}$$

$$= \frac{x^2 + 2x\ (1+2x)}{(1+2x)^{\frac{1}{2}}} \qquad \text{(Note that } (1 + 2x)^{\frac{1}{2}} \times (1 + 2x)^{\frac{1}{2}} = (1 + 2x)^{\frac{1}{2} + \frac{1}{2}} = 1 + 2x)$$

$$= \frac{x^2 + 2x + 4x^2}{(1+2x)^{\frac{1}{2}}}$$

$$\frac{dy}{dx} = \frac{5x^2 + 2x}{\sqrt{1+2x}}$$

4. Find the derivative of $(2x + 3)^3(4x^2 - 1)^2$

Solution

$$y = (2x + 3)^3(4x^2 - 1)^2$$

Let $u = (2x + 3)^3$

and $v = (4x^2 - 1)^2$

$\dfrac{du}{dx} = 3(2x + 3)^{3-1} \times 2$ (Use of chain rule. Also note that 2 is from the derivative of $2x + 3$)

$= 6(2x + 3)^2$

$\dfrac{dv}{dx} = 2(4x^2 - 1)^{2-1} \times 8x$ (Note that $8x$ is from the derivative of $4x^2 - 1$)

$= 16x(4x^2 - 1)$

Hence, $\dfrac{dy}{dx} = u\dfrac{dv}{dx} + v\dfrac{du}{dx}$

$= (2x + 3)^3 \times 16x(4x^2 - 1) + (4x^2 - 1)^2 \times 6(2x + 3)^2$

Let us factorize the expression by taking out $(2x + 3)^2$ and $(4x^2 - 1)$ which are the common terms as follows:

$\dfrac{dy}{dx} = (2x + 3)^2(4x^2 - 1)[(2x + 3)16x + (4x^2 - 1)6]$

$= (2x + 3)^2(4x^2 - 1)(32x^2 + 48x + 24x^2 - 6)$

$= (2x + 3)^2(4x^2 - 1)(56x^2 + 48x - 6)$

$= (2x + 3)^2(4x^2 - 1)\,2(28x^2 + 24x - 3)$

$\dfrac{dy}{dx} = 2(2x + 3)^2(4x^2 - 1)(28x^2 + 24x - 3)$

5. Differentiate: $y = x(x + 1)(x^2 - 4)$

Solution

$y = x(x + 1)(x^2 - 4)$

Expanding the first two brackets gives:

$y = (x^2 + x)(x^2 - 4)$

Hence, $u = (x^2 + x)$

and $v = (x^2 - 4)$

Hence, $\dfrac{du}{dx} = 2x + 1$

$\dfrac{dv}{dx} = 2x$

Therefore, $\dfrac{dy}{dx} = u\dfrac{dv}{dx} + v\dfrac{du}{dx}$

$= (x^2 + x)2x + (x^2 - 4)(2x + 1)$

$= 2x^3 + 2x^2 + 2x^3 + x^2 - 8x - 4$

$\dfrac{dy}{dx} = 4x^3 + 3x^2 - 8x - 4$

6. Find the derivative of $(x^2 + 3x - 2)^2 \sqrt{x}$

Solution

$y = (x^2 + 3x - 2)^2 \sqrt{x}$

$$u = (x^2 + 3x - 2)^2$$

$$v = \sqrt{x}$$

Or, $v = x^{\frac{1}{2}}$

$$\frac{du}{dx} = 2(x^2 + 3x - 2)^{2-1} \times 2x + 3 \quad \text{(Note that } 2x + 3 \text{ is from the derivative of } x^2 + 3x - 2)$$

$$= (4x + 6)(x^2 + 3x - 2)$$

$$\frac{dv}{dx} = \frac{1}{2}x^{-\frac{1}{2}} \quad \text{(Note that } 8x \text{ is from the derivative of } 4x^2 - 1)$$

$$= \frac{1}{2x^{\frac{1}{2}}}$$

$$\frac{dv}{dx} = \frac{1}{2\sqrt{x}}$$

Hence, $\dfrac{dy}{dx} = u\dfrac{dv}{dx} + v\dfrac{du}{dx}$

$$= (x^2 + 3x - 2)^2 \frac{1}{2\sqrt{x}} + \sqrt{x}(4x + 6)(x^2 + 3x - 2)$$

$$= (x^2 + 3x - 2)^2 \frac{1}{2\sqrt{x}} + 2\sqrt{x}(2x + 3)(x^2 + 3x - 2)$$

Factorize the expression by taking out $(x^2 + 3x - 2)$ which is the common factor. This gives:

$$\frac{dy}{dx} = (x^2 + 3x - 2)\left[\frac{x^2 + 3x - 2}{2\sqrt{x}} + 2\sqrt{x}(2x + 3)\right]$$

Simplifying the part in the bracket by taking $2\sqrt{x}$ as the LCM gives:

$$\frac{dy}{dx} = (x^2 + 3x - 2)\left[\frac{x^2 + 3x - 2 + 4x(2x + 3)}{2\sqrt{x}}\right] \quad \text{(Note that } 2\sqrt{x} \times 2\sqrt{x} = 4x)$$

$$= (x^2 + 3x - 2)\left[\frac{x^2 + 3x - 2 + 8x^2 + 12x}{2\sqrt{x}}\right]$$

$$\frac{dy}{dx} = \frac{(x^2 + 3x - 2)(9x^2 + 15x - 2)}{2\sqrt{x}}$$

7. Find the derivative of $(1 + x)(5x - 2)^{\frac{3}{2}}$

Solution

$$y = (1 + x)(5x - 2)^{\frac{3}{2}}$$

$$u = (1 + x)$$

$$v = (5x - 2)^{\frac{3}{2}}$$

$$\frac{du}{dx} = 1$$

$$\frac{dv}{dx} = \frac{3}{2}(5x - 2)^{\frac{3}{2} - 1} \times 5 \quad \text{(Note that the derivative of } 5x - 2 \text{ is 5)}$$

$$= \frac{15}{2}(5x - 2)^{\frac{1}{2}}$$

Hence, $\dfrac{dy}{dx} = u\dfrac{dv}{dx} + v\dfrac{du}{dx}$

$$= (1+x) \times \frac{15}{2}(5x-2)^{\frac{1}{2}} + (5x-2)^{\frac{3}{2}} \times 1$$

$$= \frac{15}{2}(1+x)(5x-2)^{\frac{1}{2}} + (5x-2)^{\frac{3}{2}}$$

Factorize by taking out $(5x-2)^{\frac{1}{2}}$ (i.e. the lower exponent) which is the common factor gives:

$$\frac{dy}{dx} = (5x-2)^{\frac{1}{2}}\left[\frac{15}{2}(1+x)+5x-2\right] \quad \text{(Note that } \frac{(5x-2)^{\frac{3}{2}}}{(5x-2)^{\frac{1}{2}}} = (5x-2)^{\frac{3}{2}-\frac{1}{2}} = 5x-2\text{)}$$

$$= (5x-2)^{\frac{1}{2}}\left[\frac{15}{2}+\frac{15x}{2}+5x-2\right]$$

$$= (5x-2)^{\frac{1}{2}}\left[\frac{25x}{2}+\frac{11}{2}\right]$$

$$\frac{dy}{dx} = \sqrt{5x-2}\left[\frac{25x+11}{2}\right]$$

8. If $y = (1+x)(2-3x)(2x-1)$, find $\frac{dy}{dx}$ by using product rule.

Solution

$$y = (1+x)(2-3x)(2x-1)$$

This is a product of three expressions, u, v and w.

Hence, $u = (1+x)$

$\qquad v = (2-3x)$

and $\quad w = (2x-1)$

Therefore, $\dfrac{du}{dx} = 1$

$\dfrac{dv}{dx} = -3$

$\dfrac{dw}{dx} = 2$

Hence the formula for product rule of three terms is given by:

$$\frac{dy}{dx} = \frac{du}{dx}vw + \frac{dv}{dx}uw + \frac{dw}{dx}uv$$

$$= 1(2-3x)(2x-1) + (-3)(1+x)(2x-1) + 2(1+x)(2-3x)$$

$$= 4x - 2 - 6x^2 + 3x + (-3 - 3x)(2x-1) + (2+2x)(2-3x)$$

$$= 7x - 2 - 6x^2 - 6x + 3 - 6x^2 + 3x + 4 - 6x + 4x - 6x^2$$

$$= -2 + 3 + 4 + 7x - 6x + 3x - 6x + 4x - 6x^2 - 6x^2 - 6x^2$$

$$\frac{dy}{dx} = 5 + 2x - 18x^2$$

9. Differentiate with respect to x: $(x^2 - 3x + 5)(2x - 7)$

Solution

$$y = (x^2 - 3x + 5)(2x - 7)$$

Let us differentiate this product by applying product rule but without the use u and v. This is

45

done as follows:

$$\frac{dy}{dx} = (x^2 - 3x + 5)\frac{d(2x-7)}{dx} + (2x-7)\frac{d(x^2-3x+5)}{dx}$$

$$= (x^2 - 3x + 5)2 + (2x - 7)(2x - 3)$$

$$= 2x^2 - 6x + 10 + 4x^2 - 6x - 14x + 21$$

$$\frac{dy}{dx} = 6x^2 - 26x + 31$$

10. If $y = (5x^2 - 3)(2 + \frac{3}{x})$, find $\frac{dy}{dx}$

Solution

$$y = (5x^2 - 3)(2 + \frac{3}{x})$$

$$u = (5x^2 - 3)$$

$$v = (2 + \frac{3}{x})$$

$$\frac{du}{dx} = 10x$$

$$\frac{dv}{dx} = \frac{d(3x^{-1})}{dx} \qquad \text{(Note that } \frac{3}{x} = 3x^{-1})$$

$$= -3x^{-2}$$

$$\frac{dv}{dx} = \frac{-3}{x^2}$$

Hence, $\frac{dy}{dx} = u\frac{dv}{dx} + v\frac{du}{dx}$

$$= (5x^2 - 3)\left(\frac{-3}{x^2}\right) + (2 + \frac{3}{x})10x$$

$$= -15 + \frac{9}{x^2} + 20x + 30$$

$$\frac{dy}{dx} = 15 + 20x + \frac{9}{x^2}$$

Exercise 4

1. If $f(x) = (2x - 1)(3x + 1)$, find $f'(x)$

2. Find the derivative of $y = (3x^2 - 5)(x^2 + 10)$

3. If $y = 5x(3 + x)^{\frac{1}{2}}$ find $\frac{dy}{dx}$

4. Find the derivative of $(x + 5)(x^2 - 7)^3$

5. Differentiate: $y = 2x(3x + 2)(2x^2 - 5)$

6. Find the derivative of $(3x^2 - 1)^3 \sqrt{2x}$

7. Find the derivative of $(9 - x)(x + 3)^{\frac{3}{4}}$

8. If $y = (2 - x)(5 - 3x^2)(x + 3)$, find $\dfrac{dy}{dx}$ by using product rule.

9. Differentiate with respect to x: $(3x^4 - x^2 + 2x)(x - 1)$

10. If $y = (3x^2 - x)(1 + \dfrac{1}{2x})$, find $\dfrac{dy}{dx}$

11. Find the derivative of $x^2(\sqrt{x^5})$

12. Find the derivative of $x^4(2x - 11)^{\frac{2}{3}}$

13. If $y = (7 + x)(1 - x^2)(5x^3 + 1)$, find $\dfrac{dy}{dx}$.

14. Differentiate with respect to x: $(3x^4 - x^3 + x^2 + 2x - 3)(5x + 4)$

15. If $y = (x^3 - 3x + 5)\left(\dfrac{1}{x^5}\right)$, find $\dfrac{dy}{dx}$

CHAPTER 5
QUOTIENT RULE OF DERIVATIVE

If $y = \dfrac{u}{v}$ where u and v are functions of x, then:

$$\frac{dy}{dx} = \frac{v\frac{du}{dx} - u\frac{dv}{dx}}{v^2}$$

This is called the quotient rule of differentiation.

Examples

1. If $y = \dfrac{3x^2 - 8x + 5}{5x - 2}$ find $\dfrac{dy}{dx}$

<u>Solution</u>

$$y = \frac{3x^2 - 8x + 5}{5x - 2}$$

This is of the form $y = \dfrac{u}{v}$. Therefore, we are going to apply product rule.

Let $u = 3x^2 - 8x + 5$

and $v = 5x - 2$

Hence, $\dfrac{du}{dx} = 6x - 8$

$\dfrac{dv}{dx} = 5$

Therefore, $\dfrac{dy}{dx} = \dfrac{v\frac{du}{dx} - u\frac{dv}{dx}}{v^2}$

$$= \frac{(5x - 2)(6x - 8) - (3x^2 - 8x + 5)(5)}{(5x - 2)^2}$$

$$= \frac{30x^2 - 40x - 12x + 16 - 15x^2 - 40x - 25}{(5x - 2)^2}$$

$$\frac{dy}{dx} = \frac{15x^2 - 12x - 9}{(5x - 2)^2}$$

2. Differentiate with respect to x, the function: $\dfrac{3x^2 - 2x}{x + 5}$

<u>Solution</u>

$$y = \frac{3x^2 - 2x}{x + 5}$$

Let $u = 3x^2 - 2x$

and $v = x + 5$

Hence, $\dfrac{du}{dx} = 6x - 2$

$\dfrac{dv}{dx} = 1$

Therefore, $\dfrac{dy}{dx} = \dfrac{v\dfrac{du}{dx} - u\dfrac{dv}{dx}}{v^2}$

$= \dfrac{(x+5)(6x-2) - (3x^2-2x)(1)}{(x+5)^2}$

$= \dfrac{6x^2 - 2x + 30x - 10 - 3x^2 + 2x}{(x+5)^2}$

$\dfrac{dy}{dx} = \dfrac{3x^2 + 30x - 10}{(x+5)^2}$

3. Find the differential coefficient of $y = \dfrac{-3}{x^2+5}$

Solution

$y = \dfrac{-3}{x^2+5}$

$u = -3$

and $v = x^2 + 5$

Hence, $\dfrac{du}{dx} = 0$ (The derivative of a constant is zero)

$\dfrac{dv}{dx} = 2x$

Therefore, $\dfrac{dy}{dx} = \dfrac{v\dfrac{du}{dx} - u\dfrac{dv}{dx}}{v^2}$

$= \dfrac{(x^2+5) \times 0 - (-3)2x}{(x^2+5)^2}$

$= \dfrac{0 - 6x}{(x^2+5)^2}$

$\dfrac{dy}{dx} = \dfrac{-6x}{(x^2+5)^2}$

4. Differentiate: $y = \dfrac{\sqrt{3-x}}{\sqrt{3+x}}$

Solution

$y = \dfrac{\sqrt{3-x}}{\sqrt{3+x}}$

$u = \sqrt{3-x}$

$= (3-x)^{\frac{1}{2}}$

and $v = \sqrt{3+x}$

$= (3+x)^{\frac{1}{2}}$

Hence, $\dfrac{du}{dx} = \dfrac{1}{2}(3-x)^{\frac{1}{2}-1} \times -1$ (Note that the derivative of $3-x$ is -1)

$= -\dfrac{1}{2}(3-x)^{-\frac{1}{2}}$

$$\frac{du}{dx} = \frac{-(3-x)^{-\frac{1}{2}}}{2}$$

$$\frac{dv}{dx} = \frac{1}{2}(3+x)^{\frac{1}{2}-1} \text{ x } 1 \quad \text{(Note that the derivative of } 3-x \text{ is } -1\text{)}$$

$$= \frac{1}{2}(3+x)^{-\frac{1}{2}}$$

$$\frac{dv}{dx} = \frac{(3+x)^{-\frac{1}{2}}}{2}$$

Therefore, $\dfrac{dy}{dx} = \dfrac{v\frac{du}{dx} - u\frac{dv}{dx}}{v^2}$

$$= \frac{(3+x)^{\frac{1}{2}}\left(\frac{-(3-x)^{-\frac{1}{2}}}{2}\right) - (3-x)^{\frac{1}{2}}\left(\frac{(3+x)^{-\frac{1}{2}}}{2}\right)}{(3+x)^{\frac{1}{2}})^2}$$

Taking out the common terms which are terms with positive exponent in order to factorize the expression gives:

$$\frac{dy}{dx} = \frac{\left((3+x)^{\frac{1}{2}}\right)\left((3-x)^{\frac{1}{2}}\right)\left[\frac{-(3-x)^{-1}}{2} - \frac{(3+x)^{-1}}{2}\right]}{3+x}$$

Note that in order to obtain the terms in the square bracket, we subtracted the exponents of the factors from the exponents of the original term. For example, $(3+x)^{\frac{1}{2}}\left(\dfrac{-(3-x)^{-\frac{1}{2}}}{2}\right)$ divided by $\left((3+x)^{\frac{1}{2}}\right)\left((3-x)^{\frac{1}{2}}\right)$ gave $\dfrac{-(3-x)^{-1}}{2}$ since the equal terms canceled out and $-\dfrac{1}{2} - \dfrac{1}{2} = -1$, which gave the exponent of -1. Similarly, $(3-x)^{\frac{1}{2}}\left(\dfrac{(3+x)^{-\frac{1}{2}}}{2}\right)$ divided by $\left((3+x)^{\frac{1}{2}}\right)\left((3-x)^{\frac{1}{2}}\right)$ gave $\dfrac{(3+x)^{-1}}{2}$ since the equal terms canceled out and $-\dfrac{1}{2} - \dfrac{1}{2} = -1$, which gave the exponent of -1 as the terms in the square bracket.

Let us now continue with the solution by simplifying further as follows:

$$\frac{dy}{dx} = \frac{\left((3+x)^{\frac{1}{2}}\right)\left((3-x)^{\frac{1}{2}}\right)\left[\frac{-(3-x)^{-1}}{2} - \frac{(3+x)^{-1}}{2}\right]}{3+x}$$

$$= \frac{\left((3+x)^{\frac{1}{2}}\right)\left((3-x)^{\frac{1}{2}}\right)\left[\frac{-1}{2(3-x)} - \frac{1}{2(3+x)}\right]}{3+x}$$

$$= \frac{\left((3+x)^{\frac{1}{2}}\right)\left((3-x)^{\frac{1}{2}}\right)\left[\frac{-(3+x) - (3-x)}{2(3-x)(3+x)}\right]}{3+x}$$

$$= \frac{\left((3+x)^{\frac{1}{2}}\right)\left((3-x)^{\frac{1}{2}}\right)\left[\frac{-3-x-3+x}{2(3-x)(3+x)}\right]}{3+x}$$

$$= \frac{\left((3+x)^{\frac{1}{2}}\right)\left((3-x)^{\frac{1}{2}}\right)\left[\frac{-6}{2(3-x)(3+x)}\right]}{3+x}$$

$$= \frac{\left((3+x)^{\frac{1}{2}}\right)\left((3-x)^{\frac{1}{2}}\right)\left[\frac{-3}{(3-x)(3+x)}\right]}{3+x}$$

$$= \frac{-3\left((3+x)^{\frac{1}{2}}\right)\left((3-x)^{\frac{1}{2}}\right)}{(3-x)(3+x)(3+x)}$$

$$= \frac{-3\left((3+x)^{\frac{1}{2}}\right)\left((3-x)^{\frac{1}{2}}\right)}{(3+x)^2(3-x)}$$

$$= -3(3+x)^{\frac{1}{2}-2}(3-x)^{\frac{1}{2}-1} \quad \text{(Subtraction of exponents due to the division above)}$$

$$= -3(3+x)^{-\frac{3}{2}}(3-x)^{-\frac{1}{2}}$$

$$= \frac{-3}{(3+x)^{\frac{3}{2}}(3-x)^{\frac{1}{2}}}$$

$$= \frac{-3}{\sqrt{(3+x)^3(3-x)}}$$

5. Differentiate with respect to x: $y = \dfrac{(2x^2-3)^3}{x}$

Solution

$$y = \frac{(2x^2-3)^3}{x}$$
$$u = (2x^2-3)^3$$

and $v = x$

Hence, $\dfrac{du}{dx} = 3(2x^2-3)^{3-1} \times 4x$

$$= 12x(2x^2-3)^2$$

$$\frac{dv}{dx} = 1$$

Therefore, $\dfrac{dy}{dx} = \dfrac{v\frac{du}{dx} - u\frac{dv}{dx}}{v^2}$

$$= \frac{x[12x(2x^2-3)^2] - (2x^2-3)^3 \times 1}{x^2}$$

$$= \frac{12x^2(2x^2-3)^2 - (2x^2-3)^3}{x^2}$$

Factorize the expression by taking out $(2x^2-3)^2$ which has the lower exponent. This gives:

51

$$\frac{dy}{dx} = \frac{(2x^2-3)^2[12x^2-(2x^2-3)]}{x^2}$$

$$= \frac{(2x^2-3)^2(12x^2-2x^2+3)}{x^2}$$

$$\frac{dy}{dx} = \frac{(2x^2-3)^2(10x^2+3)}{x^2}$$

6. Find the derivative of $\dfrac{\sqrt{(1+2x^2)^3}}{x}$

Solution

$$y = \frac{\sqrt{(1+2x^2)^3}}{x}$$

$$u = \sqrt{(1+2x^2)^3}$$

$$= [(1+2x^2)^3]^{\frac{1}{2}} \qquad \text{(When the square root sign is removed, we use an exponent of } \tfrac{1}{2}\text{)}$$

$$u = (1+2x^2)^{\frac{3}{2}} \qquad \text{(After multiplying the exponents)}$$

$$v = x$$

Hence, $\dfrac{du}{dx} = \dfrac{3}{2}(1+2x^2)^{\frac{3}{2}-1} \times 4x$ (Note that the derivative of $3-x$ is -1)

$$= 6x(1+2x^2)^{\frac{1}{2}}$$

$$\frac{dv}{dx} = 1$$

Therefore, $\dfrac{dy}{dx} = \dfrac{v\dfrac{du}{dx} - u\dfrac{dv}{dx}}{v^2}$

$$= \frac{x\left(6x(1+2x^2)^{\frac{1}{2}}\right) - (1+2x^2)^{\frac{3}{2}} \times 1}{x^2}$$

Take out $(1+2x^2)^{\frac{1}{2}}$ which has the lower exponent and factorize the expression. This gives:

$$\frac{dy}{dx} = \frac{(1+2x^2)^{\frac{1}{2}}[6x^2 - (1+2x^2)]}{x^2} \qquad \left(\text{Note that } \frac{(1+2x^2)^{\frac{3}{2}}}{(1+2x^2)^{\frac{1}{2}}} \text{ gives } (1+2x^2) \text{ by subtracting exponents.}\right)$$

$$= \frac{(1+2x^2)^{\frac{1}{2}}(6x^2 - 1 - 2x^2)}{x^2}$$

$$= \frac{(1+2x^2)^{\frac{1}{2}}(4x^2 - 1)}{x^2}$$

$$\frac{dy}{dx} = \frac{(\sqrt{1+2x^2})(4x^2 - 1)}{x^2}$$

7. If $y = \dfrac{(4x^3 - 3x^2 + x + 1)^{\frac{1}{2}}}{(x+1)^2}$

Solution

$$y = \frac{(4x^3 - 3x^2 + x + 1)^{\frac{1}{2}}}{(x+1)^2}$$

$u = (4x^3 - 3x^2 + x + 1)^{\frac{1}{2}}$ (After multiplying the exponents)

$v = (x + 1)^2$

Hence, $\dfrac{du}{dx} = \dfrac{1}{2}(4x^3 - 3x^2 + x + 1)^{\frac{1}{2} - 1} \times (12x^2 - 6x + 1)$

$$= \frac{(12x^2 - 6x + 1)(4x^3 - 3x^2 + x + 1)^{-\frac{1}{2}}}{2}$$

$\dfrac{dv}{dx} = 2(x + 1)^{2-1}$

$\quad = 2(x + 1)$

Therefore, $\dfrac{dy}{dx} = \dfrac{v\frac{du}{dx} - u\frac{dv}{dx}}{v^2}$

$$= \frac{(x+1)^2(12x^2 - 6x + 1)(4x^3 - 3x^2 + x + 1)^{-\frac{1}{2}} - (4x^3 - 3x^2 + x + 1)^{\frac{1}{2}}[2(x+1)]}{[(x+1)^2]^2}$$

Take out the common terms with lower exponents [i.e. $(x + 1)$ and $(4x^3 - 3x^2 + x + 1)^{-\frac{1}{2}}$] and factorize the expression. This gives:

$$\frac{dy}{dx} = \frac{(x+1)(4x^3 - 3x^2 + x + 1)^{-\frac{1}{2}}[(x+1)(12x^2 - 6x + 1) - (4x^3 - 3x^2 + x + 1)2]}{[(x+1)^2]^2}$$

Remember to subtract the exponents of the factors from the exponents of the original expression when simplifying. Simplifying further, the expression above gives:

$$\frac{dy}{dx} = \frac{(x+1)(4x^3 - 3x^2 + x + 1)^{-\frac{1}{2}}[12x^3 - 6x^2 + x + 12x^2 - 6x + 1 - (8x^3 - 6x^2 + 2x + 2)]}{2(x+1)^4}$$

$$= \frac{(x+1)(4x^3 - 3x^2 + x + 1)^{-\frac{1}{2}}(12x^3 - 6x^2 + x + 12x^2 - 6x + 1 - 8x^3 + 6x^2 - 2x - 2)}{2(x+1)^4}$$

$$= \frac{(x+1)(4x^3 - 3x^2 + x + 1)^{-\frac{1}{2}}(4x^3 + 12x^2 - 7x - 1)}{2(x+1)^4}$$

$$= \frac{(4x^3 - 3x^2 + x + 1)^{-\frac{1}{2}}(4x^3 + 12x^2 - 7x - 1)}{2(x+1)^3}$$

Note that $(x + 1)$ cancels out from the numerator and denominator.

Therefore, $\dfrac{dy}{dx} = \dfrac{4x^3 + 12x^2 - 7x - 1}{2(4x^3 - 3x^2 + x + 1)^{\frac{1}{2}}(x+1)^3}$

8. Determine $\dfrac{d}{dx}\left(\dfrac{3 + 2x - x^2}{\sqrt{1+x}}\right)$

<u>Solution</u>

Let $y = \dfrac{3 + 2x - x^2}{\sqrt{1+x}}$

Hence, $u = 3 + 2x - x^2$

$\qquad \dfrac{du}{dx} = 2 - 2x$

$\qquad v = \sqrt{1 + x}$

$\qquad = (1 + x)^{\frac{1}{2}}$

$\qquad \dfrac{du}{dx} = \dfrac{1}{2}(1 + x)^{\frac{1}{2} - 1}$ x 1

$\qquad = \dfrac{1}{2}(1 + x)^{-\frac{1}{2}}$

Therefore, $\dfrac{dy}{dx} = \dfrac{v\frac{du}{dx} - u\frac{dv}{dx}}{v^2}$

$\qquad = \dfrac{(1 + x)^{\frac{1}{2}}(2 - 2x) - (3 + 2x - x^2)\frac{1}{2}(1 + x)^{-\frac{1}{2}}}{[(1 + x)^{\frac{1}{2}}]^2}$

Take out $(1 + x)^{-\frac{1}{2}}$ as the common factor since it has the lower exponent and factorize the expression. This gives:

$\dfrac{dy}{dx} = \dfrac{(1 + x)^{-\frac{1}{2}}\left[(1+x)(2 - 2x) - (3 + 2x - x^2)\frac{1}{2}\right]}{1+x}$

$\qquad = \dfrac{(1 + x)^{-\frac{1}{2}}\left[\frac{2(1+x)(2 - 2x) - (3 + 2x - x^2)}{2}\right]}{1+x}$

$\qquad = \dfrac{(1 + x)^{-\frac{1}{2}}\left[(2 + 2x)(2 - 2x) - 3 - 2x + x^2\right]}{2(1+x)}$

$\qquad = \dfrac{(1 + x)^{-\frac{1}{2}}(4 - 4x + 4x - 4x^2 - 3 - 2x + x^2)}{2(1+x)}$

$$= \frac{1- 2x - 3x^2}{2(1+x)^{\frac{1}{2}}(1+x)}$$ [When $(1+x)^{-\frac{1}{2}}$ is taken to the denominator it becomes $(1+x)^{\frac{1}{2}}$]

$$\frac{dy}{dx} = \frac{1- 2x - 3x^2}{2(1+x)^{\frac{3}{2}}}$$ (The exponents of same terms have been added together, i.e. $\frac{1}{2}+1 = \frac{3}{2}$)

Exercise 5

1. If $y = \dfrac{x^2 - 5x + 1}{x - 1}$ find $\dfrac{dy}{dx}$

2. Differentiate with respect to x, the function: $\dfrac{4x^2 - x}{2x + 3}$

3. Find the differential coefficient of $y = \dfrac{-7}{3x^2 + 1}$

4. Differentiate: $y = \dfrac{\sqrt{1+x}}{\sqrt{1-x}}$

5. Differentiate with respect to x: $y = \dfrac{(x^3 - 2)^2}{x^2}$

6. Find the derivative of $\dfrac{\sqrt{2+3x^2}}{x^3}$

7. If $y = \dfrac{(x^2 - x - 4)^{\frac{1}{3}}}{(2x+1)^2}$

8. Determine $\dfrac{d}{dx}\left(\dfrac{x - 2x^3}{\sqrt{2-x}}\right)$

9. Find the differential coefficient of $y = -\dfrac{1}{1 - 3x^2}$

10. Differentiate: $y = \dfrac{2 + x}{2 - x}$

CHAPTER 6
DERIVATIVE OF PARAMETRIC EQUATIONS

If y = f(t) and x = g(t) are two different functions of a common variable, t, then the two equations are called parametric equations. The variable t, is the parameter.

The derivative of a parametric equation such as the one stated above is obtained as follows:

$$\frac{dy}{dx} = \frac{\frac{dy}{dt}}{\frac{dx}{dt}}$$

Examples

1. If y = 5 + t^2 and x = 3 − 2t^2, find $\frac{dy}{dx}$

Solution

$$y = 5 + t^2$$

Hence, $\frac{dy}{dt}$ = 2t

$$x = 3 - 2t^2$$

Hence, $\frac{dx}{dt}$ = − 4t

Therefore, $\frac{dy}{dx} = \frac{\frac{dy}{dt}}{\frac{dx}{dt}}$

$$= \frac{2t}{-4t}$$

$$= -\frac{1}{2} \qquad \text{(t cancels out)}$$

2. Find $\frac{dy}{dx}$ of the functions below which are expressed in the parametric form.

$$x = \frac{5}{t^2} \quad \text{and} \quad y = 2t^5 - 3$$

Solution

$$x = \frac{5}{t^2}$$

$$x = 5t^{-2}$$

$$\frac{dx}{dt} = -10t^{-3}$$

$$y = 2t^5 - 3$$

$$\frac{dy}{dt} = 10t^4$$

Therefore, $\frac{dy}{dx} = \frac{\frac{dy}{dt}}{\frac{dx}{dt}}$

$$= \frac{10t^4}{-10t^{-3}}$$

$$= -t^{4-(-3)} \quad \text{(10 cancels out)}$$

$$= -t^{4+3}$$

$$= -t^7$$

3. If $V = \frac{4}{3}\pi r^3$ and $A = \pi r^2$, find $\frac{dA}{dV}$

Solution

$$V = \frac{4}{3}\pi r^3$$

$$\frac{dV}{dr} = 3\left(\frac{4}{3}\right)\pi r^2$$

$$= 4\pi r^2$$

$$A = \pi r^2$$

$$\frac{dA}{dr} = 2\pi r$$

$$\frac{dA}{dV} = \frac{\frac{dA}{dr}}{\frac{dV}{dr}}$$

$$= \frac{2\pi r}{4\pi r^2}$$

$$\frac{dA}{dV} = \frac{1}{2r}$$

4. Determine the derivative of the curve defined by the equations: $x = t^2 - 4t$ and $y = 2t^3 - 7t$.

Solution

$$y = 2t^3 - 7t$$

Hence, $\frac{dy}{dt} = 6t^2 - 7$

$$x = t^2 - 4t$$

Hence, $\frac{dx}{dt} = 2t - 4$

Therefore, $\frac{dy}{dx} = \frac{\frac{dy}{dt}}{\frac{dx}{dt}}$

$$= \frac{6t^2 - 7}{2t - 4}$$

5. Given that $v = u + at$ and $s = ut + \frac{1}{2}at^2$. Find $\frac{dv}{ds}$ if u and a are constants.

Solution

$$v = u + at$$

$$\frac{dv}{dt} = a$$

$$s = ut + \frac{1}{2}at^2$$

$$\frac{ds}{dt} = u + (2 \times \frac{1}{2} \times at)$$

$$= u + at$$

Hence, $\dfrac{dv}{ds} = \dfrac{\frac{dv}{dt}}{\frac{ds}{dt}}$

$$= \frac{a}{u+at}$$

6. If $y = \dfrac{t}{t-2}$ and $x = \dfrac{1}{t+1}$, find $\dfrac{dy}{dx}$

<u>Solution</u>

$$y = \frac{t}{t-2}$$

$$\frac{dy}{dt} = \frac{(t-2)(1) - t(1)}{(t-2)^2} \qquad \text{(Use of quotient rule where u = t and v = t – 2)}$$

$$= \frac{t-2-t}{(t-2)^2}$$

$$\frac{dy}{dt} = \frac{-2}{(t-2)^2}$$

$$x = \frac{1}{t+1}$$

$$\frac{dx}{dt} = \frac{(t+1)(0) - 1(1)}{(t+1)^2} \qquad \text{(Use of quotient rule where u = 1 and v = t + 1)}$$

$$= \frac{-1}{(t+1)^2}$$

Hence, $\dfrac{dy}{dx} = \dfrac{\frac{dy}{dt}}{\frac{dx}{dt}}$

$$= \frac{\frac{-2}{(t-2)^2}}{\frac{-1}{(t+1)^2}}$$

$$= \frac{-2}{(t-2)^2} \times \frac{(t+1)^2}{-1}$$

$$\frac{dy}{dx} = \frac{2(t+1)^2}{(t-2)^2}$$

7. The parametric equations of the motion of a stone are: $y = 12 + 3t - 2t^2$ and $x = 5t$. Find $\dfrac{dy}{dx}$.

<u>Solution</u>

$$y = 12 + 3t - 2t^2$$

$$\frac{dy}{dt} = 3 - 4t$$

$$x = 5t$$

$$\frac{dx}{dt} = 5$$

$$\frac{dy}{dx} = \frac{\frac{dy}{dt}}{\frac{dx}{dt}}$$

$$= \frac{3 - 4t}{5}$$

8. If the parametric equations of a parabola are $y = \frac{2mt^2}{1+t^2}$ and $x = \frac{2m}{1+t^2}$ where m is a constant, find $\frac{dy}{dx}$.

Solution

$$y = \frac{2mt^2}{1+t^2}$$

$$\frac{dy}{dt} = \frac{(1+t^2)(4mt) - 2mt^2(2t)}{(1+t^2)^2} \qquad \text{(Use of quotient rule where u = } 2mt^2 \text{ and v = } 1 + t^2)$$

$$= \frac{4mt + 4mt^3 - 4mt^3}{(1+t^2)^2}$$

$$\frac{dy}{dt} = \frac{4mt}{(1+t^2)^2}$$

$$x = \frac{2m}{1+t^2}$$

$$\frac{dx}{dt} = \frac{(1+t^2)(0) - 2m(2t)}{(1+t^2)^2} \qquad \text{(Use of quotient rule where u = 1 and v = t + 1)}$$

$$= \frac{-4mt}{(1+t^2)^2}$$

Hence, $\frac{dy}{dx} = \frac{\frac{dy}{dt}}{\frac{dx}{dt}}$

$$= \frac{\frac{4mt}{(1+t^2)^2}}{\frac{-4mt}{(1+t^2)^2}}$$

$$= \frac{4mt}{(1+t^2)^2} \times \frac{(1+t^2)^2}{-4mt}$$

$$\frac{dy}{dx} = -1 \qquad \text{(Same terms cancels out)}$$

Exercise 6

1. If $y = t^3 - 5$ and $x = t^2 + 1$ find $\dfrac{dy}{dx}$

2. Find $\dfrac{dy}{dx}$ of the functions below which are expressed in the parametric form:

$$x = \frac{1}{4 - t^3} \text{ and } y = 3 + t^2$$

3. If $V = \dfrac{1}{3}\pi r^2$ and $A = \pi r l + \pi r^2$, find $\dfrac{dV}{dA}$

4. Find the derivative of the curve defined by the equations: $x = 5t^2 - t + 3$ and $y = t^2 - t - 1$.

5. Given that $F = \dfrac{m(v - u)}{t}$ and $s = \dfrac{(u+v)t}{2}$. Find $\dfrac{dF}{ds}$ if u, v and m are constants.

6. If $y = \dfrac{5}{t^2 - 1}$ and $x = \dfrac{3}{t^4 + 1}$, find $\dfrac{dy}{dx}$

7. The parametric equations of the motion of a stone are: $y = t - 3t^4$ and $x = t^2 + 3$. Find $\dfrac{dy}{dx}$.

8. If the parametric equations of a parabola are $y = \dfrac{at^3}{2}$ and $x = \dfrac{at}{5}$ where a is a constant, find $\dfrac{dy}{dx}$.

9. If $V = \dfrac{1}{3}\pi r^3$ and $A = 4\pi r^2$, find $\dfrac{dA}{dV}$

10. Given that $G = 2m + s^2$ and $H = m^2 - \dfrac{3}{s}$. Find $\dfrac{dG}{dH}$ if m is a constant.

CHAPTER 7
DERIVATIVE OF IMPLICIT FUNCTIONS

In the function y = f(x), y is said to be expressed explicitly in terms of x. However, in expressions such as $2xy - x^2y = 5$, the relationship between y and x is said to be implicit.

In order to differentiate implicit functions, y is differentiated just like x but with the addition of $\frac{dy}{dx}$ along with the value obtained.

Examples

1. Differentiate implicitly, the expression: $2x^2 + y^2 = 9$.

Solution

$$2x^2 + y^2 = 9$$

Follow the rule of differentiation and add $\frac{dy}{dx}$ to the value obtained whenever you differentiate y. Hence we differentiate each term in the expression above as follows:

$$\frac{d(2x^2)}{dx} + \frac{d(2y^2)}{dx} = \frac{d(9)}{dx}$$

$$4x + 2y\frac{dy}{dx} = 0 \quad \text{(The derivative of } y^2 \text{ is 2y and the addition of } \frac{dy}{dx} \text{ to 2y gives } 2y\frac{dy}{dx})$$

We now make $\frac{dy}{dx}$ the subject of the formula as follows:

$$2y\frac{dy}{dx} = -4x$$

$$\frac{dy}{dx} = \frac{-4x}{2y} \quad \text{(When both sides are divided by 2y)}$$

$$\frac{dy}{dx} = \frac{-2x}{y}$$

2. If $x^3 + y^3 = 18xy$, find $\frac{dy}{dx}$

Solution

$$x^3 + y^3 = 18xy$$

$$3x^2 + 3y^2\frac{dy}{dx} = (18x \times 1\frac{dy}{dx}) + (y \times 18)$$

Note that $18xy$ is differentiated by using product rule where $18x$ is taken as u while y is take as v. Also, the derivative of y is what gave us $1\frac{dy}{dx}$. The above differentiation now simplifies to:

$$3x^2 + 3y^2\frac{dy}{dx} = 18x\frac{dy}{dx} + 18y$$

Collect terms in $\frac{dy}{dx}$ on one side in order to make $\frac{dy}{dx}$ the subject of the formula as follows:

$$3y^2\frac{dy}{dx} - 18x\frac{dy}{dx} = 18y - 3x^2$$

Factorizing the left hand side gives:

$$\frac{dy}{dx}(3y^2 - 18x) = 18y - 3x^2$$

Divide both sides by $3y^2 - 18x$. This gives:

$$\frac{dy}{dx} = \frac{18y - 3x^2}{3y^2 - 18x}$$

$$= \frac{3(6y - x^2)}{3(y^2 - 6x)}$$

$$\frac{dy}{dx} = \frac{6y - x^2}{y^2 - 6x} \qquad \text{(3 cancels out)}$$

3. Find $\frac{dy}{dx}$ given that $x^2y^2 - 3xy + 4xy^3 = 4$

<u>Solution</u>

$$x^2y^2 - 3xy + 4xy^3 = 4$$

Apply product rule to x^2y^2, $3xy$ and $4xy^3$ and differentiate appropriately as follows:

$$(x^2 \times 2y\frac{dy}{dx}) + (y^2 \times 2x) - [(3x \times 1\frac{dy}{dx}) + (y \times 3)] + (4x \times 3y^2\frac{dy}{dx}) + (y^3 \times 4) = 0$$

$$2x^2y\frac{dy}{dx} + 2xy^2 - (3x\frac{dy}{dx} + 3y) + 12xy^2\frac{dy}{dx} + 4y^3 = 0$$

$$2x^2y\frac{dy}{dx} + 2xy^2 - 3x\frac{dy}{dx} - 3y + 12xy^2\frac{dy}{dx} + 4y^3 = 0$$

$$2x^2y\frac{dy}{dx} - 3x\frac{dy}{dx} + 12xy^2\frac{dy}{dx} + 2xy^2 - 3y + 4y^3 = 0$$

$$2x^2y\frac{dy}{dx} - 3x\frac{dy}{dx} + 12xy^2\frac{dy}{dx} = 3y - 2xy^2 - 4y^3$$

Factorizing the left hand side gives:

$$\frac{dy}{dx}(2x^2y - 3x + 12xy^2) = 3y - 2xy^2 - 4y^3$$

$$\frac{dy}{dx} = \frac{3y - 2xy^2 - 4y^3}{2x^2y - 3x + 12xy^2}$$

4. Differentiate $x^4 + 6x^2y^2 - 5 = 0$ implicitly with respect to x.

<u>Solution</u>

$$x^4 + 6x^2y^2 - 5 = 0$$

We differentiate accordingly and apply product rule to $6x^2y^2$ as follows:

$$4x^3 + (6x^2 \times 2y\frac{dy}{dx}) + (y^2 \times 12x) - 0 = 0$$

$$4x^3 + 12x^2y\frac{dy}{dx} + 12xy^2 = 0$$

$$12x^2y\frac{dy}{dx} = -4x^3 - 12xy^2$$

$$\frac{dy}{dx} = \frac{-4x^3 - 12xy^2}{12x^2y}$$

$$= \frac{-4x(x^2 + 3y^2)}{12x^2y}$$

$$\frac{dy}{dx} = \frac{-(x^2 + 3y^2)}{3xy} \qquad \text{(4 and } x \text{ cancels out)}$$

5. Find $\frac{dy}{dx}$ if $\frac{x^2}{16} + \frac{y^2}{25} = 1$

<u>Solution</u>

$$\frac{x^2}{16} + \frac{y^2}{25} = 1$$

$$\frac{2x}{16} + \frac{2y}{25}\frac{dy}{dx} = 0$$

$$\frac{2y}{25}\frac{dy}{dx} = -\frac{2x}{16}$$

$$\frac{2y}{25}\frac{dy}{dx} = -\frac{x}{8}$$

$$\frac{dy}{dx} = \frac{-\frac{x}{8}}{\frac{2y}{25}}$$

$$= -\frac{x}{8} \times \frac{25}{2y}$$

$$\frac{dy}{dx} = -\frac{25x}{16y}$$

Exercise 7

1. Differentiate implicitly, the expression: $5x^3 + 3y = y^2$.

2. If $2x^2 + 3y^3 = 10$, find $\frac{dy}{dx}$

3. Find $\frac{dy}{dx}$ given that $xy - 3xy^2 + 4x^3 = 4$

4. Differentiate $2x^2 + xy^2 - 5 = x^3$ implicitly with respect to x.

5. Find $\frac{dy}{dx}$ if $\frac{x^3}{3} - \frac{y^2}{2} = 0$

6. Differentiate implicitly, the expression: $2x^2y + 5y = 1$.

7. If $2x^2y + 4y^3 = 2y$, find $\frac{dy}{dx}$

8. Find $\frac{dy}{dx}$ given that $5y^2 - 4xy = y$

63

9. Differentiate $x + y^2 - x^2y^2 = 7$ implicitly with respect to x.

10. Find $\dfrac{dy}{dx}$ if $x^5 - \dfrac{3}{y} = 2$

CHAPTER 8
DERIVATIVE OF TRIGONOMETRIC FUNCTIONS

The derivatives of trigonometric functions are as given below:

If y = sin$x$, then $\dfrac{dy}{dx}$ = cos$x$

If y = cos$x$, then $\dfrac{dy}{dx}$ = $-$ sin$x$

If y = tan$x$, then $\dfrac{dy}{dx}$ = sec^2x

If y = cot$x$, then $\dfrac{dy}{dx}$ = $-$ cosec^2x

If y = sec$x$, then $\dfrac{dy}{dx}$ = sec$x$tan$x$

If y = cosec$x$, then $\dfrac{dy}{dx}$ = $-$ cot$x$cosec$x$

Example

1. Find the derivative of cos5x

<u>Solution</u>

Let y = cos5x

We have to use chain rule from composite function due to 5x which is a function of x.

Hence, let u = 5x

Therefore, y = cosu   (By replacing 5x with u)

$\dfrac{du}{dx}$ = 5

$\dfrac{dy}{du}$ = $-$ sinu   (Recall that the derivative of cos$x$ is $-$sin$x$)

Therefore, $\dfrac{dy}{dx}$ = $\dfrac{dy}{du}$ x $\dfrac{du}{dx}$ (Chain rule)

    = $-$ sinu x 5

    = $-$ 5sinu

$\dfrac{dy}{dx}$ = $-$ 5sin5x (Since u = 5x)

2. If y = sin$\dfrac{1}{2}x$, find $\dfrac{dy}{dx}$

<u>Solution</u>

y = sin$\dfrac{1}{2}x$

Let u = $\dfrac{1}{2}x$

Therefore, y = sinu   (By replacing $\dfrac{1}{2}x$ with u)

$$\frac{du}{dx} = \frac{1}{2}$$

$$\frac{dy}{du} = \cos u \quad \text{(Recall that the derivative of } \sin x \text{ is } \cos x\text{)}$$

Therefore, $\frac{dy}{dx} = \frac{dy}{du} \times \frac{du}{dx}$ (Chain rule)

$$= \cos u \times \frac{1}{2}$$

$$= \frac{1}{2}\cos u$$

$$\frac{dy}{dx} = \frac{1}{2}\cos \frac{1}{2}x \quad \text{(Since } u = \frac{1}{2}x\text{)}$$

3. Find the derivative of $5\cos 3x$

<u>Solution</u>

$$y = 5\cos 3x$$

Let $u = 3x$

Therefore, $y = 5\cos u$

$$\frac{du}{dx} = 3$$

$$\frac{dy}{du} = 5(-\sin u) \quad \text{(Note that the constant term i.e. 5 should be used to multiply the derivative)}$$

$$= -5\sin u$$

Therefore, $\frac{dy}{dx} = \frac{dy}{du} \times \frac{du}{dx}$

$$= -5\sin u \times 3$$

$$= -15\sin u$$

$$\frac{dy}{dx} = -15\sin 3x \quad \text{(Since } u = 3x\text{)}$$

4. Find the derivative of $\sin^2 x$

<u>Solution</u>

$$y = \sin^2 x$$

Note that $\sin^2 x = \sin x \times \sin x$

Hence, let $u = \sin x$

Therefore, $y = u^2$ (i.e. $u \times u$ from $\sin x \times \sin x$)

$$\frac{du}{dx} = \cos x$$

$$\frac{dy}{du} = 2u$$

Therefore, $\frac{dy}{dx} = \frac{dy}{du} \times \frac{du}{dx}$

$$= 2u \times \cos x$$

$$= 2u\cos x$$

$$\frac{dy}{dx} = 2\sin x\cos x \quad \text{(Since } u = \sin x\text{)}$$

5. Differentiate with respect to x: $y = \tan 2x$

<u>Solution</u>

$$y = \tan 2x$$

$$\frac{dy}{dx} = \sec^2 2x \times \frac{d(2x)}{dx} \quad \text{(Note that the derivative of } \tan x \text{ is } \sec^2 x\text{)}$$

$$= \sec^2 2x \times 2$$

$$= 2\sec^2 2x$$

6. If $y = \cos^3 6x^2$, differentiate y with respect to x.

<u>Solution</u>

$$y = \cos^3 6x^2$$

Let us solve this problem without the use of v as follows:

Let $u = \cos 6x^2$

Hence, $y = u^3$

$$\frac{du}{dx} = 12x(-\sin 6x^2) \quad \text{(Note that } 12x \text{ is from the derivative of } 6x^2\text{)}$$

$$= -12x\sin 6x^2$$

$$\frac{dy}{du} = 3u^2$$

Therefore, $\dfrac{dy}{dx} = \dfrac{dy}{du} \times \dfrac{du}{dx}$

$$= 3u^2 \times (-12x\sin 6x^2)$$

$$= -36xu^2\sin 6x^2$$

$$= -36x(\cos 6x^2)^2\sin 6x^2 \quad \text{(u has been replaced with } \cos 6x^2\text{)}$$

$$\frac{dy}{dx} = -36x\cos^2 6x^2\sin 6x^2$$

7. Find the derivative of $y = \sec 3x$

<u>Solution</u>

$$y = \sec 3x$$

$$\frac{dy}{dx} = \sec 3x\tan 3x \times \frac{d(3x)}{dx} \quad \text{(Note that the derivative of } \sec x \text{ is } \sec x\tan x\text{)}$$

$$= \sec 3x\tan 3x \times 3$$

$$\frac{dy}{dx} = 3\sec 3x\tan 3x$$

8. Find the derivative of $\operatorname{cosec} 4x^3$

<u>Solution</u>

$$y = \text{cosec}4x^3$$

$$\frac{dy}{dx} = -\text{cosec}4x^3\text{cot}4x^3 \times \frac{d(4x^3)}{dx}$$ (Note that the derivative of cosec is −cosec cot)

$$= -\text{cosec}4x^3\text{cot}4x^3 \times 12x^2$$

$$= -12x^2\text{cosec}4x^3\text{cot}4x^3$$

9. Find the derivative of $\text{cot}^2 2x^4$

<u>Solution</u>

$$y = \text{cot}^2 2x^4$$

This can also be written as: $y = \text{cot}2x^4 \times \text{cot}2x^4$

Let $u = 2x^4$ (Take the function of x)

Also, let $v = \text{cot }u$ (Take a function of u without taking the exponent)

Hence, $y = v^2$ (Since $v^2 = (\text{cot }u)^2 = (\text{cot}2x^4)^2 = \text{cot}^2 2x^4$. Hence, $y = v^2$)

$$\frac{du}{dx} = 8x^3$$

$$\frac{dv}{du} = -\text{cosec}^2 u$$

$$\frac{dy}{dv} = 2v$$

Therefore, $\dfrac{dy}{dx} = \dfrac{dy}{dv} \times \dfrac{dv}{du} \times \dfrac{du}{dx}$

$$= 2v \times -\text{cosec}^2 u \times 8x^3$$

$$= -16x^3 v \,\text{cosec}^2 u$$

$$= -16x^3 \text{cot }u\,\text{cosec}^2 u$$ (Since v = cot u)

Substituting in the original value of u gives:

$$\frac{dy}{dx} = -16x^3\text{cot }2x^4\text{cosec}^2 2x^4$$

10. Find the derivative of $5x\sin 2x$

<u>Solution</u>

$$y = 5x\sin 2x$$

We are going to apply the product rule of differentiation.

Let $u = 5x$

and $v = \sin 2x$

$$\frac{du}{dx} = 5$$

$$\frac{dv}{dx} = 2\cos 2x$$ (Note that the differentiation of $2x$ gives 2)

Hence, $\dfrac{dy}{dx} = u\dfrac{dv}{dx} + v\dfrac{du}{dx}$

$$= (5x \times 2\cos 2x) + (\sin 2x \times 5)$$

$$= 10x\cos 2x + 5\sin 2x$$

$$\frac{dy}{dx} = 5(2x\cos 2x + \sin 2x)$$

11. Find $\frac{dy}{dx}$ if $y = \frac{1}{x} \sec x$

Solution

$$y = \frac{1}{x} \sec x$$

We apply product rule as follows:

$$u = \frac{1}{x}$$

$$= x^{-1}$$

$$v = \sec 3x$$

$$\frac{du}{dx} = -1x^{-2}$$

$$= \frac{-1}{x^2}$$

$$\frac{dv}{dx} = \sec 3x \tan 3x \times 3$$

$$= 3\sec 3x \tan 3x$$

Note that the derivative of sec x is sec$x$tan$x$ and 3 is from the derivative of $3x$

Hence, $\frac{dy}{dx} = u\frac{dv}{dx} + v\frac{du}{dx}$

$$= \frac{1}{x}(3\sec 3x \tan 3x) + \sec 3x \left(\frac{-1}{x^2}\right)$$

$$\frac{dy}{dx} = \frac{3}{x}\sec 3x \tan 3x - \frac{1}{x^2}\sec 3x$$

12. Find the derivative of $3\csc x^6$

Solution

$$y = 3\csc x^6$$

Let $u = x^6$

Hence, $y = 3\csc u$

$$\frac{du}{dx} = 6x^5$$

$$\frac{dy}{du} = -3\cot u \csc u \quad \text{(The constant term i.e. 5 should be used to multiply the derivative)}$$

Therefore, $\frac{dy}{dx} = \frac{dy}{du} \times \frac{du}{dx}$

$$= -3\cot u \csc u \times 6x^5$$

$$= -18x^5 \cot u \csc u$$

$$\frac{dy}{dx} = -18x^5 \cot x^6 \csc x^6 \quad \text{(Since } u = x^6)$$

13. If $y = \dfrac{1 + \cos 2x}{\sin 2x}$ find $\dfrac{dy}{dx}$.

Solution

$$y = \frac{1 + \cos 2x}{\sin 2x}$$

We have to apply quotient rule on this as follows:

$u = 1 + \cos 2x$

$v = \sin 2x$

$\dfrac{du}{dx} = -2\sin 2x$

$\dfrac{dv}{dx} = 2\cos 2x$

Hence, $\dfrac{dy}{dx} = \dfrac{v\frac{du}{dx} - u\frac{dv}{dx}}{v^2}$ (Quotient rule)

$$= \frac{\sin 2x(-2\sin 2x) - (1 + \cos 2x)(2\cos 2x)}{(\sin 2x)^2}$$

$$= \frac{-2\sin^2 2x - (2\cos 2x + 2\cos^2 2x)}{\sin^2 2x}$$ (Note that $(\sin 2x)^2 = \sin^2 2x$)

$$= \frac{-2\sin^2 2x - 2\cos 2x - 2\cos^2 2x}{\sin^2 2x}$$

$$= \frac{-2\sin^2 2x - 2\cos^2 2x - 2\cos 2x}{\sin^2 2x}$$

$$= \frac{-2(\sin^2 2x + \cos^2 2x) - 2\cos 2x}{\sin^2 2x}$$

$$= \frac{-2(1) - 2\cos 2x}{\sin^2 2x}$$ (Note that $\sin^2 x + \cos^2 x = 1$, hence $\sin^2 2x + \cos^2 2x = 1$)

$$\frac{dy}{dx} = \frac{-2(1 + \cos 2x)}{\sin^2 2x}$$

14. Differentiate $y = \dfrac{\sin^2 x}{x}$

Solution

$$y = \frac{\sin^2 x}{x}$$

This is also quotient rule.

Therefore, $u = \sin^2 x$

$v = x$

Let us follow a direct and systematic way of differentiating trigonometric functions.

Hence, in order to differentiate $\sin 2x$:

First, differentiate the exponent of $\sin^2$ without changing the trigonometric term. This gives:

$2\sin^{2-1}$

$= 2\sin$

Then add the term in x. This gives:

$2\sin x$

The next step is to differentiate sin which gives cos. Also add the term in x to obtain $\cos x$.

Finally, differentiate the term in x. Hence, differentiating x gives 1.

Now multiply the three terms obtained in the three steps above. This gives:

$2\sin x$ x $\cos x$ x 1

$= 2\sin x\cos x$

Hence, $\dfrac{du}{dx} = 2\sin x\cos x$ (As obtained above)

Since, $v = x$

Then, $\dfrac{dv}{dx} = 1$

Hence, $\dfrac{dy}{dx} = \dfrac{v\frac{du}{dx} - u\frac{dv}{dx}}{v^2}$

$= \dfrac{x(2\sin x\cos x) - \sin^2 x(1)}{x^2}$

$= \dfrac{2x\sin x\cos x - \sin^2 x}{x^2}$

$= \dfrac{\sin x(2x\cos x - \sin x)}{x^2}$

15. Differentiate with respect to x: $\dfrac{\sec 2x}{x^3 + 1}$

Solution

$y = \dfrac{\sec 2x}{x^3 + 1}$

$u = \sec 2x$

$v = x^3 + 1$

Hence, $\dfrac{du}{dx} = 2\sec 2x\tan 2x$

Then, $\dfrac{dv}{dx} = 3x^2$

Hence, $\dfrac{dy}{dx} = \dfrac{v\frac{du}{dx} - u\frac{dv}{dx}}{v^2}$

$= \dfrac{(x^3 + 1)(2\sec 2x\tan 2x) - \sec 2x(3x^2)}{(x^3 + 1)^2}$

$= \dfrac{(x^3 + 1)(2\sec 2x\tan 2x) - 3x^2\sec 2x}{(x^3 + 1)^2}$

16. Find the derivative of $\dfrac{\cos \sqrt{x}}{1 + x}$

<u>Solution</u>

$$y = \frac{\cos \sqrt{x}}{1+x}$$

$$u = \cos\sqrt{x}$$

$$= \cos x^{\frac{1}{2}}$$

$$v = 1+x$$

Hence, $\dfrac{du}{dx} = \dfrac{d(x^{\frac{1}{2}})}{dx} \text{ x } \dfrac{d(\cos)}{dx}$

Note that $\dfrac{d(\cos)}{dx}$ means the derivative of cos which gives $-\sin$, and then $x^{\frac{1}{2}}$ is added to it to give $-\sin x^{\frac{1}{2}}$. Hence we continue as follows:

$$\frac{du}{dx} = \frac{1}{2}x^{-\frac{1}{2}} \text{ x } -\sin x^{\frac{1}{2}}$$

$$= \frac{1}{2x^{\frac{1}{2}}} \text{ x } -\sin x^{\frac{1}{2}}$$

$$\frac{du}{dx} = \frac{-\sin x^{\frac{1}{2}}}{2x^{\frac{1}{2}}}$$

Also, $\dfrac{dv}{dx} = 1$

Hence, $\dfrac{dy}{dx} = \dfrac{v\dfrac{du}{dx} - u\dfrac{dv}{dx}}{v^2}$

$$= \frac{(1+x)\left(\dfrac{-\sin x^{\frac{1}{2}}}{2x^{\frac{1}{2}}}\right) - \cos x^{\frac{1}{2}}(1)}{(1+x)^2}$$

$$= \frac{-\sin \sqrt{x}}{2\sqrt{x}(1+x)} - \frac{\cos \sqrt{x}}{(1+x)^2} \quad \text{(When the fractions are separated)}$$

$$\frac{dy}{dx} = \frac{-(1+x)\sin \sqrt{x} - 2\sqrt{x}\cos \sqrt{x}}{2\sqrt{x}(1+x)^2} \quad \text{(When the fractions are combined)}$$

17. Find $\dfrac{dy}{dx}$ if $y = \dfrac{1-x^2}{1+\cos x}$

<u>Solution</u>

$$y = \frac{1-x^2}{1+\cos x}$$

$$u = 1 - x^2$$

$$v = 1 + \cos x$$

$$\frac{du}{dx} = -2x$$

$$\frac{dv}{dx} = -\sin x$$

72

Hence, $\dfrac{dy}{dx} = \dfrac{v\frac{du}{dx} - u\frac{dv}{dx}}{v^2}$

$= \dfrac{(1+\cos x)(-2x) - (1-x^2)(-\sin x)}{(1+\cos x)^2}$

$\dfrac{dy}{dx} = \dfrac{-2x(1+\cos x) + (1-x^2)\sin x}{(1+\cos x)^2}$

Note that the negative sign from $-\sin x$ changed the negative sign at the middle to a positive sign since negative sign multiplied by negative sign gives a positive sign.

18. Find the derivative of $8x\sin x^2$

Solution

$y = 8x\sin x^2$

We apply product rule as follows:

$u = 8x$

$\dfrac{du}{dx} = 8$

$v = \sin x^2$

$\dfrac{dv}{dx} = \dfrac{d(\sin)}{dx} \ \text{x} \ \dfrac{d(x^2)}{dx}$

$= \cos x^2 \ \text{x} \ 2x$

$= 2x\cos x^2$

Note that $\dfrac{d(\sin)}{dx}$ gives cos which result to $\cos x^2$ when x^2 from the question is added

Hence, $\dfrac{dy}{dx} = u\dfrac{dv}{dx} + v\dfrac{du}{dx}$

$= 8x(2x\cos x^2) + \sin x^2(8)$

$= 16x^2\cos x^2 + 8\sin x^2$

$\dfrac{dy}{dx} = 8(2x^2\cos x^2 + \sin x^2)$

19. Differentiate with respect to x: $\sec 2x\sin^3 2x$

Solution

$y = \sec 2x\sin^3 2x$

We apply product rule as follows:

$u = \sec 2x$

$\dfrac{du}{dx} = \dfrac{d(\sec)}{dx} \ \text{x} \ \dfrac{d(2x)}{dx}$

$= \sec 2x\tan 2x \ \text{x} \ 2$

$\dfrac{du}{dx} = 2\sec 2x\tan 2x$

Note that $\dfrac{d(\sec)}{dx}$ gives sectan, but remember to add $2x$ after sec and tan respectively to obtain sec$2x$tan$2x$

$v = \sin^3 2x$

In order to directly differentiate a trigonometric term with exponent like this (i.e. $\sin^3$) we have three solutions to multiply as follows:

First solution: consider only the exponent and differentiate $\sin^3$. This gives:

$$\frac{d(\sin^3)}{dx} = 3\sin^2$$

We now add $2x$ from the question to obtain $3\sin^2 2x$

Second solution: $\dfrac{d(\sin)}{dx} = \cos$ which gives cos$2x$

Third solution: $\dfrac{d(2x)}{dx} = 2$

Multiply theses three solutions to give the derivative of $v = \sin^3 2x$ as follows:

$$\frac{dv}{dx} = 3\sin^2 2x \times \cos 2x \times 2$$

$$= 6\sin^2 2x\cos 2x$$

Now, $\dfrac{dy}{dx} = u\dfrac{dv}{dx} + v\dfrac{du}{dx}$

$= \sec 2x(6\sin^2 2x\cos 2x) + \sin^3 2x(2\sec 2x\tan 2x)$

$= \dfrac{1}{\cos 2x}(6\sin^2 2x\cos 2x) + \sin^3 2x(2\dfrac{1}{\cos 2x}\dfrac{\sin 2x}{\cos 2x})$ (Note that $\sec 2x = \dfrac{1}{\cos 2x}$ and $\tan 2x = \dfrac{\sin 2x}{\cos 2x}$)

$= 6\sin^2 2x + 2\sin^2 2x(\dfrac{\sin 2x}{\cos 2x}\dfrac{\sin 2x}{\cos 2x})$

Note that one $\sin 2x$ has been taken out of $\sin^3 2x$ and placed inside the bracket.

$= 6\sin^2 2x + 2\sin^2 2x(\tan 2x\tan x)$ (Since $\dfrac{\sin 2x}{\cos 2x} = \tan 2x$)

$= 6\sin^2 2x + 2\sin^2 2x(\tan^2 2x)$

$\dfrac{dy}{dx} = 2\sin^2 2x(3 + \tan^2 2x)$ (After factorization)

20. Differentiate with respect to x: $\dfrac{\sec x - \tan x}{\sec x + \tan x}$

Solution

$y = \dfrac{\sec x - \tan x}{\sec x + \tan x}$

$u = \sec x - \tan x$

$\dfrac{du}{dx} = \sec x\tan x - \sec^2 x$

$v = \sec x + \tan x$

$\dfrac{dv}{dx} = \sec x\tan x + \sec^2 x$

$$\frac{dy}{dx} = \frac{v\frac{du}{dx} - u\frac{dv}{dx}}{v^2}$$

$$= \frac{(\sec x + \tan x)(\sec x\tan x - \sec 2x) - (\sec x - \tan x)(\sec x\tan x + \sec 2x)}{(\sec x + \tan x)^2}$$

Expanding bracket in the numerator gives:

$$\frac{dy}{dx} = \frac{\sec^2 x\tan x - \sec^3 x + \sec x\tan^2 x - \sec^2 x\tan x - [\sec^2 x\tan x + \sec^3 x - \sec x\tan^2 x - \sec^2 x\tan x]}{(\sec x + \tan x)^2}$$

$$= \frac{\sec^2 x\tan x - \sec^3 x + \sec x\tan^2 x - \sec^2 x\tan x - \sec^2 x\tan x - \sec^3 x + \sec x\tan^2 x + \sec^2 x\tan x]}{(\sec x + \tan x)^2}$$

$$= \frac{2\sec x\tan^2 x - 2\sec^3 x}{(\sec x + \tan x)^2} \qquad \text{(Note that all the sec}^2 x\text{tan}x \text{ have cancelled out)}$$

$$= \frac{2\sec x(\tan^2 x - \sec^2 x)}{(\sec x + \tan x)^2}$$

$$= \frac{2\sec x(-1)}{(\sec x + \tan x)^2} \qquad \text{(Note that tan}^2 x - \text{sec}^2 x = -1)$$

$$= \frac{-2\sec x}{(\sec x + \tan x)^2}$$

$$\frac{dy}{dx} = -\frac{2\sec x}{(\sec x + \tan x)^2}$$

21. Find the derivative of $\sqrt{\dfrac{\cos 2x}{1 + \sin 2x}}$

Solution

$$y = \sqrt{\frac{\cos 2x}{1 + \sin 2x}}$$

This can also be written as:

$$y = \frac{(\cos 2x)^{\frac{1}{2}}}{(1 + \sin 2x)^{\frac{1}{2}}}$$

$$u = (\cos 2x)^{\frac{1}{2}}$$

Using the chain rule, we obtain $\dfrac{du}{dx}$ as follows:

$$\frac{du}{dx} = \frac{1}{2}(\cos 2x)^{\frac{1}{2} - 1} \times \frac{d(\cos 2x)}{dx}$$

$$= \frac{1}{2}(\cos 2x)^{-\frac{1}{2}} \times -2\sin 2x$$

$$= \frac{-2\sin 2x}{2(\cos 2x)^{\frac{1}{2}}}$$

$$= \frac{-\sin 2x}{(\cos 2x)^{\frac{1}{2}}}$$

$$v = (1 + \sin 2x)^{\frac{1}{2}}$$

75

$$\frac{dv}{dx} = \frac{1}{2}(1 + \sin2x)^{\frac{1}{2}-1} \times \frac{d(1+\sin 2x)}{dx}$$

$$= \frac{1}{2}(1 + \sin2x)^{-\frac{1}{2}} \times 2\cos2x$$

$$= \frac{2\cos 2x}{2(1 + \sin 2x)^{\frac{1}{2}}}$$

$$= \frac{\cos 2x}{(1 + \sin 2x)^{\frac{1}{2}}}$$

$$\frac{dy}{dx} = \frac{v\frac{du}{dx} - u\frac{dv}{dx}}{v^2}$$

$$= \frac{(1 + \sin 2x)^{\frac{1}{2}}\left(\dfrac{-\sin 2x}{(\cos 2x)^{\frac{1}{2}}}\right) - (\cos 2x)^{\frac{1}{2}}\left(\dfrac{\cos 2x}{(1 + \sin 2x)^{\frac{1}{2}}}\right)}{[(1 + \sin 2x)^{\frac{1}{2}}]^2}$$

$$= \frac{\left(\dfrac{-\sin 2x(1 + \sin 2x)^{\frac{1}{2}}}{(\cos 2x)^{\frac{1}{2}}}\right) - \left(\dfrac{\cos 2x(\cos 2x)^{\frac{1}{2}}}{(1 + \sin 2x)^{\frac{1}{2}}}\right)}{1 + \sin 2x}$$

$$= \frac{\left(\dfrac{-\sin 2x(1 + \sin 2x) - \cos 2x(\cos 2x)}{(\cos 2x)^{\frac{1}{2}}(1 + \sin 2x)^{\frac{1}{2}}}\right)}{1 + \sin 2x}$$

Note that $(\cos2x)^{\frac{1}{2}}$ x $(\cos2x)^{\frac{1}{2}}$ = $\cos2x$ (Since the exponents are added). Similarly:

$(1 + \sin2x)^{\frac{1}{2}}$ x $(1 + \sin2x)^{\frac{1}{2}}$ = $1 + \sin2x$. Simplifying the above expression further, gives:

$$\frac{dy}{dx} = \frac{-\sin 2x(1 + \sin 2x) - \cos^2 2x}{(\cos 2x)^{\frac{1}{2}}(1 + \sin 2x)^{\frac{1}{2}}(1 + \sin2x)}$$

$$= \frac{-\sin 2x - \sin^2 2x - \cos^2 2x}{(\cos 2x)^{\frac{1}{2}}(1 + \sin 2x)^{\frac{1}{2}}(1 + \sin2x)}$$

$$= \frac{-\sin 2x - (\sin^2 2x + \cos^2 2x)}{(\cos 2x)^{\frac{1}{2}}(1 + \sin 2x)^{\frac{1}{2}}(1 + \sin2x)}$$

$$= \frac{-\sin 2x - 1}{(\cos 2x)^{\frac{1}{2}}(1 + \sin 2x)^{\frac{1}{2}}(1 + \sin2x)} \qquad \text{(Note that } \sin^2 2x + \cos^2 2x = 1)$$

$$= \frac{-(1 + \sin 2x)}{(\cos 2x)^{\frac{1}{2}}(1 + \sin 2x)^{\frac{1}{2}}(1 + \sin2x)}$$

$$= \frac{-1}{(\cos 2x)^{\frac{1}{2}}(1 + \sin 2x)^{\frac{1}{2}}} \qquad \text{(Note that } 1 + \sin2x \text{ cancels out)}$$

22. Differentiate with respect to x: $\sin x - 2x\cos x$

Solution

$y = \sin x - 2x\cos x$

Treat $2x\cos x$ using product rule.

$$\frac{dy}{dx} = \frac{d(\sin x)}{dx} - \left(2x\,\frac{d(\cos x)}{dx} + \cos x\,\frac{d(2x)}{dx}\right)$$

$$= \cos x - [2x(-\sin x) + \cos x(2)]$$

$$= \cos x + 2x\sin x - 2\cos x$$

$$= 2x\sin x - \cos x$$

23. Find the derivative of $\cos^5 3x^4$

Solution

$y = \cos^5 3x^4$

A direct way of differentiating this problem is applied as follows:

$$\frac{dy}{dx} = \frac{d(\cos^5)}{dx} \times \frac{d(\cos)}{dx} \times \frac{d(3x^4)}{dx}$$

$$= 5\cos^4 3x^4 \times (-\sin 3x^4) \times 12x^3$$

Note that the derivative of $\cos^5$ gives $5\cos^4$ (do not change the trigonometric term, i.e. cos) and then the addition of $3x^4$ from the question gives $5\cos^4 3x^4$.

Similarly the derivative of cos gives $-\sin$ and the addition of $3x^4$ from the question gives $-\sin 3x^4$.

Hence, multiplying the terms above gives:

$$\frac{dy}{dx} = -60x^3\cos^4 3x^4 \sin 3x^4$$

24. Differentiate with respect to x: $\sin^8 15x^6$

Solution

$y = \sin^8 15x^6$

We can also differentiate this problem directly as follows:

$$\frac{dy}{dx} = \frac{d(\sin^8)}{dx} \times \frac{d(\sin)}{dx} \times \frac{d(15x^6)}{dx}$$

$$= 8\sin^7 15x^6 \times \cos 15x^6 \times 90x^5$$

Note that the derivative of $\sin^8$ gives $8\sin^7$ (in this case do not change the trigonometric term, i.e. sin) and then the addition of $15x^6$ from the question gives $8\sin^7 15x^6$.

Similarly the derivative of sin gives cos and the addition of $15x^6$ from the question gives $\cos 15x^6$. Hence, we continue as follows:

$$\frac{dy}{dx} = 8\sin^7 15x^6 \times \cos 15x^6 \times 90x^5$$

$$= 720x^5\sin^7 15x^6\cos 15x^6$$

Exercise 8

1. Find the derivative of $\tan 2x$

2. If $y = \cos\dfrac{1}{5}x$, find $\dfrac{dy}{dx}$

3. Find the derivative of $10\cos 5x$

4. Find the derivative of $\sin^3 x$

5. Differentiate with respect to x: $y = \tan^2 x$

6. If $y = \sin^4 3x^5$, differentiate y with respect to x.

7. Find the derivative of $y = \operatorname{cosec} 6x^2$

8. Find the derivative of $\sec 2x^5$

9. Find the derivative of $\tan 3x^2$

10. Find the derivative of $x^2\cos 3x$

11. Find $\dfrac{dy}{dx}$ if $y = \dfrac{3}{x^3}\sin 2x$

12. Find the derivative of $12\sec x^4$

13. If $y = \dfrac{2 - \sin 5x}{\tan x}$ find $\dfrac{dy}{dx}$.

14. Differentiate $y = \dfrac{\sin^3 x}{2x}$

15. Differentiate with respect to x: $\dfrac{\cot 2x}{x + 3}$

16. Find the derivative of $\dfrac{\sin\sqrt[3]{3}}{5x}$

17. Find $\dfrac{dy}{dx}$ if $y = \dfrac{x^2 - 3}{\sec 2x}$

18. Find the derivative of $3x^2\cos 3x^2$

19. Differentiate with respect to x: $\cot x\cos^2 x$

20. Differentiate with respect to x: $\dfrac{\sin x - \cos x}{\sin x + \cos x}$

21. Find the derivative of $\dfrac{\sec 4x}{\cos 4x - 2}$

22. Differentiate with respect to x: $\cos 3x - x^2\sec x$

23. Find the derivative of $\sin^9 x^3$

24. Differentiate with respect to x: $\sin 6x^{\frac{1}{2}}$

25. Find $\dfrac{dy}{dx}$ if $y = \dfrac{1}{x^2}\sin x^3$

26. Find the derivative of $\cos^3 2x^5$

27. If $y = \dfrac{\cos 10x}{\sin 2x}$ find $\dfrac{dy}{dx}$.

28. Differentiate $y = \dfrac{3\sin x^4}{2x}$

29. Differentiate with respect to x: $\dfrac{\tan 5x}{2x - 1}$

30. Find the derivative of $\dfrac{\tan^2 x}{2x}$

CHAPTER 9
DERIVATIVE OF INVERSE FUNCTIONS

If the derivative of a function is given by $\dfrac{dy}{dx}$, then the derivative of the inverse function is given

by: $\quad \dfrac{1}{\frac{dy}{dx}} = \dfrac{dx}{dy}$

Or, $\quad \dfrac{dy}{dx} = \dfrac{1}{\frac{dx}{dy}}$

Examples

1. Find $\dfrac{dx}{dy}$ if y = $\sqrt[3]{x}$

Solution

Method 1

$$y = \sqrt[3]{x}$$

Or $y = x^{\frac{1}{3}}$

Let us make x the subject of the formula. The inverse of $\dfrac{1}{3}$ is 3. Hence raise both sides to the

exponent 3 as follows:

$$y^3 = (x^{\frac{1}{3}})^3$$

$$y^3 = x^1 \qquad \text{(Note that 1 was obtained from } \dfrac{1}{3} \times 3)$$

Hence, $x = y^3$

Therefore, $\dfrac{dx}{dy} = 3y^2$

Method 2

$$y = \sqrt[3]{x}$$

Or $y = x^{\frac{1}{3}}$

$$\dfrac{dy}{dx} = \dfrac{1}{3} x^{\frac{1}{3}-1}$$

$$= \dfrac{1}{3} x^{-\frac{2}{3}}$$

$$\dfrac{dy}{dx} = \dfrac{1}{3x^{\frac{2}{3}}}$$

Hence, $\dfrac{dx}{dy} = \dfrac{1}{\frac{dy}{dx}}$

$$= \dfrac{3x^{\frac{2}{3}}}{1} \qquad \text{(This means the inverse of } \dfrac{1}{3x^{\frac{2}{3}}})$$

$= 3x^{\frac{2}{3}}$

$= 3(\sqrt[3]{x})^2$ [Recall from indices that $x^{\frac{a}{b}} = (\sqrt[b]{x})^a$]

$\dfrac{dx}{dy} = 3y^2$ (Since $y = \sqrt[3]{x}$)

2. If $y = \sqrt[5]{2x - 3}$ find $\dfrac{dx}{dy}$

Solution

$y = \sqrt[5]{2x - 3}$

Or $y = (2x - 3)^{\frac{1}{5}}$

Let us make x the subject of the formula. The inverse of $\dfrac{1}{5}$ is 5. Hence raise both sides to the exponent 5 as follows:

$y^5 = [(2x - 3)^{\frac{1}{5}}]^5$

$y^5 = 2x - 3$ (Note that $\dfrac{1}{5}$ x 5 = 1, which cancels the fractional exponent)

Hence, $2x = y^5 + 3$

$x = \dfrac{y^5 + 3}{2}$

$x = \dfrac{y^5}{2} + \dfrac{3}{2}$ (When we separate into fractions by dividing each part by the denominator)

Therefore, $\dfrac{dx}{dy} = \dfrac{5y^4}{2}$

3. If $y = x^3 - 5$. Find $\dfrac{dx}{dy}$

Solution

$y = x^3 - 5$

$y + 5 = x^3$

$x^3 = y + 5$

$x = (y + 5)^{\frac{1}{3}}$ (By raising both sides to an exponent of the inverse of 3 which is $\dfrac{1}{3}$)

$\dfrac{dx}{dy} = \dfrac{1}{3}(y + 5)^{\frac{1}{3} - 1}$ x 1

Note that the 1 was obtained from the derivative of y + 5 since chain rule was used.

$\dfrac{dx}{dy} = \dfrac{1}{3}(y + 5)^{-\frac{2}{3}}$

$\dfrac{dx}{dy} = \dfrac{1}{3(y+5)^{\frac{2}{3}}}$

Recall from indices that $x^{\frac{a}{b}} = (\sqrt[b]{x})^a$. Applying this rule gives:

$$\frac{dx}{dy} = \frac{1}{3[\sqrt[3]{(y+5)^2}]}$$

4. If $y = \frac{1}{2}x^4 + 3$, find $\frac{dx}{dy}$

Solution

$$y = \frac{1}{2}x^4 + 3$$

$$y - 3 = \frac{1}{2}x^4$$

$$2(y - 3) = x^4$$

$$x^4 = 2y - 6$$

$$x = (2y - 6)^{\frac{1}{4}} \quad \text{(This is obtained by raising both sided to an exponent of the inverse of 4, i.e. } \frac{1}{4})$$

Hence we use chain rule to determine $\frac{dx}{dy}$ as follows:

$$\frac{dx}{dy} = \frac{1}{4}(2y - 6)^{\frac{1}{4} - 1} \times 2 \qquad \text{(Note that 2 is from the derivative of } 2y - 6)$$

$$\frac{dx}{dy} = \frac{1}{2}(2y - 6)^{-\frac{3}{4}}$$

$$\frac{dx}{dy} = \frac{1}{2(2y - 6)^{\frac{3}{4}}}$$

Hence, $\frac{dx}{dy} = \frac{1}{2[\sqrt[4]{(2y-6)^3}]}$ \qquad [This is obtained from the law of indices given by: $x^{\frac{a}{b}} = (\sqrt[b]{x})^a$]

5. If $y = \frac{x + 2}{x}$, find $\frac{dx}{dy}$.

Solution

$$y = \frac{x + 2}{x}$$

$$xy = x + 2 \qquad \text{(When we cross multiply)}$$

$$xy - x = 2$$

Factorizing the left hand side gives:

$$x(y - 1) = 2$$

$$x = \frac{2}{y - 1}$$

$$x = 2(y - 1)^{-1} \qquad \text{(Take note of the use of negative exponent when the denominator goes up)}$$

Hence we use chain rule to determine $\frac{dx}{dy}$ as follows:

$$\frac{dx}{dy} = -1 \times 2(y - 1)^{-1-1} \times 1 \qquad \text{(Note that 1 is from the derivative of } y - 1)$$

$$\frac{dx}{dy} = -2(y - 1)^{-2}$$

$$\frac{dx}{dy} = \frac{-2}{(y - 1)^2}$$

6. Find $\frac{dx}{dy}$ if $y = \frac{1}{x+2}$

Solution

$$y = \frac{1}{x+2}$$

$$x + 2 = \frac{1}{y}$$

$$x = \frac{1}{y} - 2$$

$$x = y^{-1} - 2$$

$$\frac{dx}{dy} = -1 \times y^{-1-1}$$

$$= -y^{-2}$$

$$\frac{dx}{dy} = \frac{-1}{y^2}$$

7. Find $\frac{dx}{dy}$ if $y = x^{\frac{2}{3}}$

Solution

$$y = x^{\frac{2}{3}}$$

Raise both sides to the exponent $\frac{3}{2}$ i.e. the inverse of $\frac{2}{3}$. This gives:

$$y^{\frac{3}{2}} = (x^{\frac{2}{3}})^{\frac{3}{2}}$$

$$y^{\frac{3}{2}} = x \quad \text{(Note that } \frac{2}{3} \times \frac{3}{2} = 1, \text{ and } x^1 = x)$$

$$x = y^{\frac{3}{2}}$$

$$\frac{dx}{dy} = \frac{3}{2} \times y^{\frac{3}{2}-1}$$

$$= \frac{3}{2} y^{\frac{1}{2}}$$

$$\frac{dx}{dy} = \frac{3\sqrt{y}}{2}$$

8. If $y = (5x + 7)^{\frac{3}{10}}$ find $\frac{dx}{dy}$.

Solution

$$y = (5x + 7)^{\frac{3}{10}}$$

Raise both sides to the exponent $\frac{10}{3}$ i.e. the inverse of $\frac{3}{10}$. This gives:

$$y^{\frac{10}{3}} = [(5x + 7)^{\frac{3}{10}}]^{\frac{10}{3}}$$

$$y^{\frac{10}{3}} = 5x + 7 \quad \text{(Note that } \frac{3}{10} \times \frac{10}{3} = 1)$$

$$y^{\frac{10}{3}} - 7 = 5x$$

$$x = \frac{y^{\frac{10}{3}} - 7}{5}$$

$$= \frac{y^{\frac{10}{3}}}{5} - \frac{7}{5}$$

$$x = \frac{1}{5}y^{\frac{10}{3}} - \frac{7}{5}$$

$$\frac{dx}{dy} = \frac{10}{3} \times \frac{1}{5}y^{\frac{10}{3} - 1}$$

$$= \frac{2}{3}y^{\frac{7}{3}}$$

$$\frac{dx}{dy} = \frac{2\sqrt[3]{y^7}}{3}$$

Exercise 9

1. Find $\frac{dx}{dy}$ if $y = 7x^3$

2. If $y = \sqrt[3]{1 + x^2}$ find $\frac{dx}{dy}$

3. If $y = 2x^5 - 3$. Find $\frac{dx}{dy}$

4. If $y = \frac{2}{3}x^3 - 9$, find $\frac{dx}{dy}$

5. If $y = \frac{2x + 1}{x}$, find $\frac{dx}{dy}$.

6. Find $\frac{dx}{dy}$ if $y = \frac{3}{x^2 - 5}$

7. Find $\frac{dx}{dy}$ if $y = 2x^{\frac{1}{4}}$

8. If $y = (x - 3)^{\frac{1}{5}}$ find $\frac{dx}{dy}$.

9. If $y = \dfrac{4x + 5}{x}$, find $\dfrac{dx}{dy}$.

10. Find $\dfrac{dx}{dy}$ if $y = \dfrac{1}{x^3 + 8}$

CHAPTER 10
DERIVATIVES OF INVERSE TRIGONOMETRIC FUNCTIONS

Recall that if $\sin x = y$, then $x = \sin^{-1} y$. This is referred to as inverse trigonometric function. The derivatives of inverse trigonometric functions are given below.

If $y = \sin^{-1} x$, then $\dfrac{dy}{dx} = \dfrac{1}{\sqrt{1-x^2}}$

If $y = \cos^{-1} x$, then $\dfrac{dy}{dx} = \dfrac{-1}{\sqrt{1-x^2}}$

If $y = \tan^{-1} x$, then $\dfrac{dy}{dx} = \dfrac{1}{1+x^2}$

If $y = \cot^{-1} x$, then $\dfrac{dy}{dx} = \dfrac{-1}{1+x^2}$

If $y = \sec^{-1} x$, then $\dfrac{dy}{dx} = \dfrac{1}{x\sqrt{x^2-1}}$

If $y = \operatorname{cosec}^{-1} x$, then $\dfrac{dy}{dx} = \dfrac{-1}{x\sqrt{x^2-1}}$

Note that $\sin^{-1} x$ can also be written as $\arcsin x$. Other inverse function can be written in a similar way.

Examples

1. If $y = \cos^{-1} x$ find the $\dfrac{dy}{dx}$.

<u>Solution</u>

$\quad y = \cos^{-1} x$

$\quad \dfrac{dy}{dx} = \dfrac{-1}{\sqrt{1-x^2}}$

2. Find the derivative of $\sin^{-1} 3x$

<u>Solution</u>

$\quad y = \sin^{-1} 3x$

Let us use the chain rule to solve this problem

Let $u = 3x$

Hence, $y = \sin^{-1} u$

$\quad \dfrac{du}{dx} = 3$

$\quad \dfrac{dy}{du} = \dfrac{1}{\sqrt{1-u^2}}$

Therefore, $\dfrac{dy}{dx} = \dfrac{dy}{du} \times \dfrac{du}{dx}$ (Chain rule)

$$= \frac{1}{\sqrt{1-u^2}} \times 3$$

$$= \frac{3}{\sqrt{1-u^2}}$$

$$= \frac{3}{\sqrt{1-(3x)^2}} \qquad \text{(Since } u = 3x\text{)}$$

$$\frac{dy}{dx} = \frac{3}{\sqrt{1-9x^2}}$$

3. Find the derivative of $\cot^{-1}x^2$

<u>Solution</u>

$$y = \cot^{-1}x^2$$

Let $u = x^2$

Hence, $y = \cot^{-1}u$

$$\frac{du}{dx} = 2x$$

$$\frac{dy}{du} = \frac{-1}{1+u^2}$$

Therefore, $\dfrac{dy}{dx} = \dfrac{dy}{du} \times \dfrac{du}{dx}$

$$= \frac{-1}{1+u^2} \times 2x$$

$$= \frac{-2x}{1+u^2}$$

$$= \frac{-2x}{1+(x^2)^2} \qquad \text{(since } u = x^2\text{)}$$

$$\frac{dy}{dx} = \frac{-2x}{1+x^4}$$

4. If $y = \sec^{-1}2x^3$, find $y = \dfrac{dy}{dx}$

<u>Solution</u>

$$y = \sec^{-1}2x^3$$

Let $u = 2x^3$

Hence, $y = \sec^{-1}u$

$$\frac{du}{dx} = 6x^2$$

$$\frac{dy}{du} = \frac{1}{u\sqrt{u^2-1}}$$

Therefore, $\dfrac{dy}{dx} = \dfrac{dy}{du} \times \dfrac{du}{dx}$

$$= \frac{1}{u\sqrt{u^2-1}} \times 6x^2$$

$$= \frac{6x^2}{u\sqrt{u^2-1}}$$

$$= \frac{6x^2}{2x^3\sqrt{(2x^3)^2-1}} \qquad \text{(since } u = 2x^3)$$

$$\frac{dy}{dx} = \frac{3}{x\sqrt{4x^6-1}} \qquad \text{(Note that } \frac{6x^2}{2x^3} = \frac{3}{x})$$

5. Find the derivative $x^2\tan^{-1}x$

<u>Solution</u>

$$y = x^2\tan^{-1}x$$

we apply the product rule as follows:

$$u = x^2$$

$$v = \tan^{-1}x$$

$$\frac{du}{dx} = 2x$$

$$\frac{dv}{dx} = \frac{1}{1+x^2}$$

$$\frac{dy}{dx} = u\frac{dv}{dx} + v\frac{du}{dx} \qquad \text{(product rule)}$$

$$= x^2\left(\frac{1}{1+x^2}\right) + \tan^{-1}x(2x)$$

$$\frac{dy}{dx} = \frac{x^2}{1+x^2} + 2x\tan^{-1}x$$

6. If $y = 3x - 1 \cosec^{-1}x^3$, find $\frac{dy}{dx}$.

<u>Solution</u>

$$y = 3x - 1 \cosec^{-1}x^3$$

$$u = 3x - 1$$

$$\frac{du}{dx} = 3$$

$$v = \cosec^{-1}x^3$$

$$\frac{dv}{dx} = \frac{-1}{x^3\sqrt{(x^3)^2-1}} \times 3x^2 \qquad \text{(From use of chain rule. Note that } 3x^2 \text{ is from the derivative of } x^3)$$

$$= \frac{-3x^2}{x^3\sqrt{x^6-1}}$$

$$\frac{dv}{dx} = \frac{-3}{x\sqrt{x^6-1}}$$

$$\frac{dy}{dx} = u\frac{dv}{dx} + v\frac{du}{dx} \qquad \text{(product rule)}$$

$$= 3x - 1\left(\frac{-3}{x\sqrt{x^6 - 1}}\right) + \csc^{-1}x^3(3)$$

$$\frac{dy}{dx} = \frac{-3(3x - 1)}{x\sqrt{x^6 - 1}} + 3\csc^{-1}x^3$$

7. Find the derivative of $\cos^{-1}(5x - 3)$

Solution

$$y = \cos^{-1}(5x - 3)$$

Let $u = 5x - 3$

Hence, $y = \cos^{-1}u$

$$\frac{du}{dx} = 5$$

$$\frac{dy}{du} = \frac{-1}{\sqrt{1 - u^2}}$$

Therefore, $\dfrac{dy}{dx} = \dfrac{dy}{du} \times \dfrac{du}{dx}$

$$= \frac{-1}{\sqrt{1 - u^2}} \times 5$$

$$= \frac{-5}{\sqrt{1 - u^2}}$$

$$= \frac{-5}{\sqrt{1 - (5x - 3)^2}} \qquad \text{(Since } u = 5x - 3)$$

Expanding the bracket gives:

$$\frac{dy}{dx} = \frac{-5}{\sqrt{1 - (25x^2 - 15x - 15x + 9)}}$$

$$= \frac{-5}{\sqrt{1 - 25x^2 + 15x + 15x - 9}}$$

$$\frac{dy}{dx} = \frac{-5}{\sqrt{-25x^2 + 30x - 8}}$$

8. If $y = \tan^{-1}\left(\frac{1}{m}\right)$, find $\dfrac{dy}{dm}$

Solution

$$y = \tan^{-1}\left(\frac{1}{m}\right)$$

Let $u = \dfrac{1}{m} = m^{-1}$

Hence, $y = \tan^{-1}u$

$$\frac{du}{dm} = -1m^{-2}$$

$$\frac{du}{dm} = \frac{-1}{m^2}$$

$$\frac{dy}{du} = \frac{1}{1 + u^2}$$

Therefore, $\dfrac{dy}{dm} = \dfrac{dy}{du} \times \dfrac{du}{dm}$

$= \dfrac{1}{1+u^2} \times \dfrac{-1}{m^2}$

$= \dfrac{1}{1+\left(\frac{1}{m}\right)^2} \times \dfrac{-1}{m^2}$

$= \dfrac{-1}{m^2\left[1+\left(\frac{1}{m}\right)^2\right]}$

$= \dfrac{-1}{m^2\left(1+\frac{1}{m^2}\right)}$

$\dfrac{dy}{dm} = \dfrac{-1}{m^2+1}$

9. If $y = \sin 3x$, find $\dfrac{dx}{dy}$.

Solution

$\quad y = \sin 3x$

$\quad \sin^{-1}y = 3x \qquad$ (Recall that if $\sin a = b$, then $a = \sin^{-1}b$)

$\quad x = \dfrac{\sin^{-1}y}{3}$

$\quad x = \dfrac{1}{3}\sin^{-1}y$

Hence, $\dfrac{dx}{dy} = \dfrac{1}{3}\dfrac{1}{\sqrt{1-y^2}}$

$\dfrac{dx}{dy} = \dfrac{1}{3\sqrt{1-y^2}}$

Note that this example asked us to find $\dfrac{dx}{dy}$ and not $\dfrac{dy}{dx}$

10. If $y = (\cos 5x)^2$ find $\dfrac{dx}{dy}$.

Solution

$\quad y = (\cos 5x)^2$

This can also be written as $\cos^2 5x$. Hence:

$\quad y = \cos^2 5x.$

Or, $\cos^2 5x = y$

taking the square root of both sides gives:

$\quad \cos 5x = \sqrt{y}$

$\quad 5x = \cos^{-1}\sqrt{y}$

$$x = \frac{\cos^{-1}\sqrt{y}}{5}$$

$$= \frac{1}{5}\cos^{-1}\sqrt{y}$$

$$x = \frac{1}{5}\cos^{-1}y^{\frac{1}{2}}$$

Hence, $\dfrac{dx}{dy} = \dfrac{1}{5} \cdot \dfrac{-1}{\sqrt{1-(y^{\frac{1}{2}})^2}} \times \dfrac{1}{2}y^{-\frac{1}{2}}$ (Note that $\dfrac{1}{2}y^{-\frac{1}{2}}$ is from the derivative of $y^{\frac{1}{2}}$)

$$= \frac{-1}{5\sqrt{1-y}} \times \frac{1}{2y^{\frac{1}{2}}}$$

$$= \frac{-1}{10y^{\frac{1}{2}}\sqrt{1-y}}$$

$$= \frac{-1}{10\sqrt{y}\sqrt{1-y}}$$

$$\frac{dx}{dy} = \frac{-1}{10\sqrt{y(1-y)}}$$

Exercise 10

1. If $y = \sin^{-1}2x$ find the $\dfrac{dy}{dx}$.

2. Find the derivative of $\cos^{-1}x$

3. Find the derivative of $\cot^{-1}3x^2$

4. If $y = \cot^{-1}x^4$, find $y = \dfrac{dy}{dx}$

5. Find the derivative $3x\sec^{-1}x$

6. If $y = x^2 + 3\tan^{-1}x$, find $\dfrac{dy}{dx}$.

7. Find the derivative of $\tan^{-1}(x + 5)$

8. If $y = \cos^{-1}\left(\dfrac{1}{x}\right)$, find $\dfrac{dy}{dx}$

9. If $y = \cos5x$, find $\dfrac{dx}{dy}$.

10. If $y = (\sec x)^2$ find $\dfrac{dx}{dy}$.

11. Find the derivative $5x\tan^{-1}3x$

12. If $y = x^2 + 4\sin^{-1}x^5$, find $\dfrac{dy}{dx}$.

13. Find the derivative of $\csc^{-1}x^3$

14. If y = tan2x, find $\dfrac{dx}{dy}$.

15. If y = sin^{3}5x, find $\dfrac{dx}{dy}$.

CHAPTER 11
DERIVATIVES OF HYPERBOLIC FUNCTIONS

Hyperbolic functions are functions in calculus which are expressed by the combination of the exponential functions e^x and e^{-x}. The derivatives of the six main hyperbolic functions are as given below.

1. If $y = \sinh x$, then $\dfrac{dy}{dx} = \cosh x$

2. If $y = \cosh x$, then $\dfrac{dy}{dx} = \sinh x$

3. If $y = \tanh x$, then $\dfrac{dy}{dx} = \operatorname{sech}^2 x$

4. If $y = \coth x$, then $\dfrac{dy}{dx} = -\operatorname{cosech}^2 x$

5. If $y = \operatorname{sech} x$, then $\dfrac{dy}{dx} = -\operatorname{sech} x \tanh x$

6. If $y = \operatorname{cosech} x$, then $\dfrac{dy}{dx} = -\operatorname{cosech} x \coth x$

Note that $\sinh x = \dfrac{e^x - e^{-x}}{2}$ and $\cosh x = \dfrac{e^x + e^{-x}}{2}$

Examples

1. If $y = \cosh x - 5\sinh x$, find $\dfrac{dy}{dx}$.

<u>Solution</u>

$$y = \cosh x - 5\sinh x$$

$$\dfrac{dy}{dx} = \sinh x - 5\cosh x$$

2. Find the derivative of $2x^3 \coth x$

<u>Solution</u>

$$y = 2x^3 \coth x$$

Using product rule gives $\dfrac{dy}{dx}$ as follows:

$$\dfrac{dy}{dx} = 2x^3\left[\dfrac{d(\coth x)}{dx}\right] + \coth x\left[\dfrac{d(2x^3)}{dx}\right]$$

$$= 2x^3(-\operatorname{cosech}^2 x) + \coth x(6x^2)$$

$$= -2x^3\operatorname{cosech}^2 x + 6x^2\coth x$$

$$\dfrac{dy}{dx} = -2x^2(x\operatorname{cosech}^2 x - 3\coth x)$$

Take note of the change in sign of the term in the bracket. This is due to the negative sign outside the bracket. Expanding the bracket gives the original expression that was factorized.

3. If $y = \dfrac{\cosh x}{x^2 + 1}$ find $\dfrac{dy}{dx}$.

Solution

Let us use product rule to obtained $\dfrac{dy}{dx}$ as follows:

$$\frac{dy}{dx} = \frac{(x^2+1)\left[\frac{d(\cosh x)}{dx}\right] - \cosh x \left[\frac{d(x^2+1)}{dx}\right]}{(x^2+1)^2}$$

$$= \frac{(x^2+1)(\sinh x) - \cosh x(2x)}{(x^2+1)^2}$$

$$\frac{dy}{dx} = \frac{\sinh x(x^2+1) - 2x\cosh x}{(x^2+1)^2}$$

4. Find the derivative of $(\sinh 3x)^2$

Solution

$y = (\sinh 3x)^2$ (This can also be written as $\sinh^2 3x$)

Let $u = \sinh 3x$

Hence, $y = u^2$

$\dfrac{du}{dx} = 3\cosh 3x$ (Note that 3 is from the derivative of $3x$)

$\dfrac{dy}{du} = 2u$

$\dfrac{dy}{dx} = \dfrac{dy}{du} \times \dfrac{du}{dx}$

     $= 2u \times 3\cosh 3x$

     $= 6u\cosh 3x$

     $= 6\sinh 3x \cosh 3x$ (u has been replaced with $\sinh 3x$)

5. Find the derivative of $\sinh^4 2x^3$

Solution

$y = \sinh^4 2x^3$

Let $u = \sinh 2x^3$

Hence, $y = u^4$

$\dfrac{du}{dx} = 6x^2 \cosh 2x^3$ (Note that $6x^2$ is from the derivative of $2x^3$)

$\dfrac{dy}{du} = 4u^3$

$\dfrac{dy}{dx} = \dfrac{dy}{du} \times \dfrac{du}{dx}$

     $= 4u^3 \times 6x^2 \cosh 2x^3$

     $= 24x^2 u^3 \cosh 2x^3$

$\dfrac{dy}{dx} = 24x^2 \sinh^3 2x^3 \cosh 2x^3$ (u has been replaced with $\sinh 2x^3$)

6. Find the derivative of $\text{cosech}4x^3$

Solution

$y = \text{cosech}4x^3$

$\dfrac{dy}{dx} = -\text{cosech}4x^3\text{coth}4x^3 \times \dfrac{d(4x^3)}{dx}$ (Note that the derivative of cosech is $-$cosechcoth)

$= -\text{cosech}4x^3\text{coth}4x^3 \times 12x^2$

$= -12x^2\text{cosech}4x^3\text{coth}4x^3$

7. Find the derivative of $\text{coth}^2 2x^4$

Solution

$y = \text{coth}^2 2x^4$

Let $u = \text{coth}2x^4$

Hence, $y = u^2$

$\dfrac{du}{dx} = -\text{cosech}^2 2x^4 \times \dfrac{d(2x^4)}{dx}$

$= -\text{cosech}^2 2x^4 \times 8x^3$

$\dfrac{du}{dx} = -8x^3\text{cosech}^2 2x^4$

$\dfrac{dy}{du} = 2u$

Therefore, $\dfrac{dy}{dx} = \dfrac{dy}{du} \times \dfrac{du}{dx}$

$= 2u \times (-8x^3\text{cosech}^2 2x^4)$

$= -16x^3 u \text{ cosech}^2 2x^4$

$\dfrac{dy}{dx} = -16x^3\text{coth}2x^4\text{cosech}^2 2x^4$ (Since $u = \text{coth}2x^4$)

8. Find the derivative of $5x\text{sinh}2x$

Solution

$y = 5x\text{sinh}2x$

We apply the product rule of differentiation as follows:

Let $u = 5x$

and $v = \text{sinh}2x$

$\dfrac{du}{dx} = 5$

$\dfrac{dv}{dx} = 2\cosh2x$ (Note that the differentiation of $2x$ gives 2)

Hence, $\dfrac{dy}{dx} = u\dfrac{dv}{dx} + v\dfrac{du}{dx}$

$= (5x \times 2\cosh2x) + (\text{sinh}2x \times 5)$

$= 10x\cosh2x + 5\text{sinh}2x$

$$\frac{dy}{dx} = 5(2x\cosh 2x + \sinh 2x)$$

9. If $y = \dfrac{1 + \cosh 2x}{\sinh 2x}$ find $\dfrac{dy}{dx}$.

Solution

$$y = \frac{1 + \cosh 2x}{\sinh 2x}$$

We apply quotient rule as follows:

$u = 1 + \cosh 2x$

$v = \sinh 2x$

$\dfrac{du}{dx} = 2\sinh 2x$

$\dfrac{dv}{dx} = 2\cosh 2x$

Hence, $\dfrac{dy}{dx} = \dfrac{v\frac{du}{dx} - u\frac{dv}{dx}}{v^2}$ (Quotient rule)

$$= \frac{\sinh 2x (2\sinh 2x) - (1 + \cosh 2x)(2\cosh 2x)}{(\sinh 2x)^2}$$

$$= \frac{2\sin^2 2x - (2\cosh 2x + 2\cosh^2 2x)}{\sinh^2 2x} \qquad \text{(Note that } (\sinh 2x)^2 = \sinh^2 2x)$$

$$= \frac{2\sinh^2 2x - 2\cosh 2x - 2\cosh^2 2x}{\sinh^2 2x}$$

$$= \frac{2\sinh^2 2x - 2\cosh^2 2x - 2\cosh 2x}{\sinh^2 2x}$$

$$= \frac{-2(\cosh^2 2x - \sinh^2 2x) - 2\cosh 2x}{\sinh^2 2x}$$

$$= \frac{-2(1) - 2\cosh 2x}{\sinh^2 2x} \qquad \text{(Note that } \cosh^2 2x - \sinh^2 2x = 1)$$

$$\frac{dy}{dx} = \frac{-2(1 + \cosh 2x)}{\sinh^2 2x}$$

Or, $\dfrac{dy}{dx} = \dfrac{-2(1 + \cosh 2x)}{\cosh^2 2x - 1}$ (Note that since $\cosh^2 2x - \sinh^2 2x = 1$, then $\cosh^2 2x - 1 = \sinh^2 2x$)

$$= \frac{-2(1 + \cosh 2x)}{(\cosh 2x + 1)(\cosh 2x - 1)}$$

Note that from difference of two squares we have: $a^2 - b^2 = (a + b)(a - b)$.

Hence, $\cosh^2 2x - 1$ is also a difference of two squares since 1 is also 1^2. Therefore, $\cosh^2 2x - 1 = (\cosh 2x + 1)(\cosh 2x - 1)$ as represented above. Hence the expression above becomes:

$$\frac{dy}{dx} = \frac{-2}{(\cosh 2x - 1)}$$

10. Differentiate $y = \dfrac{\sinh^2 x}{x}$

Solution

$$y = \frac{\sinh^2 x}{x}$$

We use quotient rule as follows:

Therefore, u = $\sinh^2 x$

$v = x$

Hence, $\frac{du}{dx} = 2\sinh x \cosh x$

Since, $v = x$

Then, $\frac{dv}{dx} = 1$

Hence, $\frac{dy}{dx} = \frac{v\frac{du}{dx} - u\frac{dv}{dx}}{v^2}$

$$= \frac{x(2\sinh x \cosh x) - \sinh^2 x(1)}{x^2}$$

$$= \frac{2x\sinh x \cosh x - \sin^2 x}{x^2}$$

$$= \frac{\sinh x(2x\cosh x - \sinh x)}{x^2}$$

Exercise 11

1. If y = sinh3x – 2cosh$x$, find $\frac{dy}{dx}$.

2. Find the derivative of $5x^2$sech$x$

3. If y = $\frac{\sinh 2x}{3x^2}$ find $\frac{dy}{dx}$.

4. Find the derivative of $\cosh^2 3x$

5. Find the derivative of $\cosh^3 4x^5$

6. Find the derivative of coth$2x^5$

7. Find the derivative of $(\text{sech}2x^2)^3$

8. Find the derivative of x^2cosh5x

9. If y = $\frac{\tanh 5x}{\coth 3x}$ find $\frac{dy}{dx}$.

10. Differentiate y = $\frac{2\cosh^3 x}{3x}$

CHAPTER 12
DERIVATIVE OF LOGARITHMIC FUNCTIONS

If $y = \log_a x$, then $\dfrac{dy}{dx} = \dfrac{1}{x}\log_a e$, where a is any base.

If $y = \log_e x$, then $\dfrac{dy}{dx} = \dfrac{1}{x}$

Note that $\log_e x$ can also be represented as $\ln x$ and the value of e is 2.718 (to 3 decimal places)

Examples

1. Find the derivative of $\log_a 2x$.

Solution

$y = \log_a 2x$

We will use chain rule since we have a function that is not just x but $2x$.

Let u = 2x

Therefore, $y = \log_a u$

$\dfrac{du}{dx} = 2$

$\dfrac{dy}{du} = \dfrac{1}{u}\log_a e$

$\dfrac{dy}{dx} = \dfrac{dy}{du} \times \dfrac{du}{dx}$

$\quad = \dfrac{1}{u}\log_a e \times 2$

$\quad = \dfrac{2}{u}\log_a e$

$\quad = \dfrac{2}{2x}\log_a e \qquad$ (since u = 2x)

$\dfrac{dy}{dx} = \dfrac{1}{x}\log_a e$

2. Find the derivative of $\log_a(5x - 1)$

Solution

$y = \log_a(5x - 1)$

Let u = 5x − 1

Therefore, $y = \log_a u$

$\dfrac{du}{dx} = 5$

$\dfrac{dy}{du} = \dfrac{1}{u}\log_a e$

$\dfrac{dy}{dx} = \dfrac{dy}{du} \times \dfrac{du}{dx}$

$$= \frac{1}{u} \log_a e \times 5$$

$$= \frac{5}{u} \log_a e$$

$$\frac{dy}{dx} = \frac{5}{5x-1} \log_a e \qquad \text{(since } u = 5x - 1\text{)}$$

3. Differentiate $\log_a(4x-3)^2$ with respect to x

Solution

$$y = \log_a(4x-3)^2$$

Let $u = (4x-3)^2$

Therefore, $y = \log_a u$

$$\frac{du}{dx} = 2(4x-3)^{2-1} \times 4 \qquad \text{(Note that 4 is from the derivative of } 4x - 3\text{)}$$

$$\frac{du}{dx} = 8(4x-3)$$

$$\frac{dy}{du} = \frac{1}{u} \log_a e$$

$$\frac{dy}{dx} = \frac{dy}{du} \times \frac{du}{dx}$$

$$= \frac{1}{u} \log_a e \times 8(4x-3)$$

$$= \frac{8(4x-3)}{u} \log_a e$$

$$= \frac{8(4x-3)}{(4x-3)^2} \log_a e \qquad [\text{since } u = (4x-3)^2]$$

$$\frac{dy}{dx} = \frac{8}{4x-3} \log_a e \qquad \text{(One } 4x - 3 \text{ cancels out)}$$

4. If $y = \log_a \sqrt{1 + 2x}$, find $\dfrac{dy}{dx}$

Solution

$$y = \log_a \sqrt{1 + 2x}$$

Let $u = \sqrt{1+2x} = (1+2x)^{\frac{1}{2}}$

Therefore, $y = \log_a u$

$$\frac{du}{dx} = \frac{1}{2}(1+2x)^{\frac{1}{2}-1} \times 2$$

$$= (1+2x)^{-\frac{1}{2}} \qquad \text{(Note that } \frac{1}{2} \times 2 = 1\text{)}$$

$$\frac{du}{dx} = \frac{1}{(1+2x)^{\frac{1}{2}}}$$

$$\frac{dy}{du} = \frac{1}{u} \log_a e$$

$$\frac{dy}{dx} = \frac{dy}{du} \times \frac{du}{dx}$$

$$= \frac{1}{u}\log_a e \times \frac{1}{(1+2x)^{\frac{1}{2}}}$$

$$= \frac{1}{(1+2x)^{\frac{1}{2}}}\log_a e \times \frac{1}{(1+2x)^{\frac{1}{2}}} \qquad \text{[Note that } u \text{ has been replaced with } (1+2x)^{\frac{1}{2}}]$$

$$\frac{dy}{dx} = \frac{1}{1+2x}\log_a e \qquad \text{[Note that } (1+2x)^{\frac{1}{2}} \times (1+2x)^{\frac{1}{2}} = (1+2x)^1 \text{ by adding exponents]}$$

5. Find $\frac{dy}{dx}$ given that y = $\log_a\frac{1-3x}{1+3x}$

<u>Solution</u>

$$y = \log_a\frac{1-3x}{1+3x}$$

Or, y = $\log_a(1-3x) - \log_a(1+3x)$ \qquad (Recall that $\log_x\left(\frac{a}{b}\right) = \log_x a - \log_x b$)

$$\frac{dy}{dx} = \frac{-3}{1-3x}\log_a e - \frac{3}{1+3x}\log_a e$$

$$= -\log_a e\left(\frac{3}{1-3x} + \frac{3}{1+3x}\right)$$

$$= -\log_a e\left[\frac{3(1+3x) + 3(1-3x)}{(1-3x)(1+3x)}\right]$$

$$= -\log_a e\left[\frac{3+9x+3-9x}{1-9x^2}\right] \qquad \text{[Note that } (1-3x)(1+3x) = 1-9x^2]$$

$$= -\log_a e\left(\frac{6}{1-9x^2}\right)$$

$$\frac{dy}{dx} = \frac{-6}{1-9x^2}\log_a e$$

6. If y = $\log_{10}(x^2 - 2)$, find $\frac{dy}{dx}$.

<u>Solution</u>

$$y = \log_{10}(x^2 - 2)$$

The value 10 represents a in other examples. So we are going to differentiate y like the examples above except that we will write 10 wherever 'a' should be.

$$y = \log_{10}(x^2 - 2)$$
$$u = x^2 - 2$$

Hence, y = $\log_{10}u$

$$\frac{du}{dx} = 2x$$

$$\frac{dy}{du} = \frac{1}{u}\log_{10}e \qquad \text{(Just like differentiating } \log_a u\text{)}$$

$$\frac{dy}{dx} = \frac{dy}{du} \times \frac{du}{dx}$$

$$= \frac{1}{u} \log_{10}e \times 2x$$

$$= \frac{2x}{u} \log_{10}e$$

$$\frac{dy}{dx} = \frac{2x}{x^2 - 2} \log_{10}e$$

7. Find $\frac{dy}{dx}$ if $y = \log_{10}\frac{1}{x}$

Solution

$$y = \log_{10}\frac{1}{x}$$

Let us solve this question directly without using $u = \frac{1}{x}$ as follows:

$$\frac{dy}{dx} = \frac{\frac{d\left(\frac{1}{x}\right)}{dx}}{\frac{1}{x}} \times \log_{10}e$$

$$= \frac{\frac{d(x^{-1})}{dx}}{x^{-1}} \times \log_{10}e$$

$$= \frac{-x^{-2}}{x^{-1}} \times \log_{10}e \qquad \text{(Note that the derivative of } x^{-1} \text{ is } -x^{-2})$$

$$= -x^{-2} \times x^{1} \times \log_{10}e \qquad \text{(Since } \frac{1}{x^{-1}} = x^{1})$$

$$= -x^{-1} \times \log_{10}e \qquad \text{(After adding the exponents of } x)$$

$$\frac{dy}{dx} = -\frac{1}{x} \log_{10}e$$

8. Find the derivative of $\log_e(2 - x^3)$

Solution

$$y = \log_e(2 - x^3)$$

Note that the base here is 'e' and not 'a'.

Let $u = 2 - x^3$

Hence, $y = \log_e u$

$$\frac{du}{dx} = -3x^2$$

$$\frac{dy}{du} = \frac{1}{u} \qquad \text{(Recall that the derivative of } \log_e x = \frac{1}{x})$$

$$\frac{dy}{dx} = \frac{dy}{du} \times \frac{du}{dx}$$

$$= \frac{1}{u} \times -3x^2$$

$$= \frac{-3x^2}{u}$$

$$\frac{dy}{dx} = \frac{-3x^2}{2 - x^3}$$

9. Find the derivative of $(\log_e 5x)^2$

Solution

$$y = (\log_e 5x)^2$$

Let $u = \log_e 5x$

Hence, $y = u^2$

$$\frac{du}{dx} = \frac{\frac{d(5x)}{dx}}{5x}$$

$$= \frac{5}{5x}$$

$$\frac{du}{dx} = \frac{1}{x}$$

$$\frac{dy}{du} = 2u$$

$$\frac{dy}{dx} = \frac{dy}{du} \times \frac{du}{dx}$$

$$= 2u \times \frac{1}{x}$$

$$= \frac{2u}{x}$$

$$\frac{dy}{dx} = \frac{2}{x} \log_e 5x \quad \text{(Since } u = \log_e 5x)$$

10. If $y = \ln\sqrt{3x^2 - 4}$, find $\frac{dy}{dx}$.

Solution

$$y = \ln\sqrt{3x^2 - 4}$$

Note that $\ln\sqrt{3x^2 - 4}$ is also the same as $\log_e\sqrt{3x^2 - 4}$

Hence, $y = \log_e\sqrt{3x^2 - 4}$

Or, $y = \log_e(3x^2 - 4)^{\frac{1}{2}}$

Let $u = (3x^2 - 4)^{\frac{1}{2}}$

Hence, $y = \log_e u$

$$\frac{du}{dx} = \frac{1}{2}(3x^2 - 4)^{\frac{1}{2}-1} \times 6x \quad \text{(Note that } 6x \text{ is from the derivative of } 3x^2 - 4)$$

$$= \frac{6x}{2}(3x^2 - 4)^{-\frac{1}{2}}$$

$$\frac{du}{dx} = \frac{3x}{(3x^2 - 4)^{\frac{1}{2}}}$$

$$\frac{dy}{du} = \frac{1}{u}$$

$$\frac{dy}{dx} = \frac{dy}{du} \times \frac{du}{dx}$$

$$= \frac{1}{u} \times \frac{3x}{(3x^2-4)^{\frac{1}{2}}}$$

$$= \frac{1}{(3x^2-4)^{\frac{1}{2}}} \times \frac{3x}{(3x^2-4)^{\frac{1}{2}}}$$

$$\frac{dy}{dx} = \frac{3x}{3x^2-4} \quad \text{(Note that } (3x^2 - 4)^{\frac{1}{2}} \times (3x^2 - 4)^{\frac{1}{2}} = 3x^2 - 4 \text{, after adding their exponents)}$$

11. Find $\dfrac{dy}{dx}$ if $y = \sqrt[3]{x}\ \ln 2x$.

Solution

$y = \sqrt[3]{x}\ \ln 2x$

We are going to apply product rule here.

$u = \sqrt[3]{x} = x^{\frac{1}{3}}$

$v = \ln 2x$ (Note that $\ln 2x$ is the same as $\log_e 2x$)

$$\frac{du}{dx} = \frac{1}{3} x^{\frac{1}{3}-1}$$

$$= \frac{1}{3} x^{-\frac{2}{3}}$$

$$\frac{dv}{dx} = \frac{2}{2x}$$

$$= \frac{1}{x}$$

Hence, $\dfrac{dy}{dx} = u\dfrac{dv}{dx} + v\dfrac{du}{dx}$ (Product rule)

$$= x^{\frac{1}{3}}\left(\frac{1}{x}\right) + \ln 2x \left(\frac{1}{3} x^{-\frac{2}{3}}\right)$$

$$= x^{\frac{1}{3}}(x^{-1}) + \frac{1}{3}x^{-\frac{2}{3}} \ln 2x$$

$$= x^{-\frac{2}{3}}\left(1 + \frac{\ln 2x}{3}\right)$$

$$= x^{-\frac{2}{3}}\left(\frac{3 + \ln 2x}{3}\right)$$

$$\frac{dy}{dx} = \frac{1}{x^{\frac{2}{3}}}\left(\frac{3 + \ln 2x}{3}\right)$$

Or, $\dfrac{dy}{dx} = \dfrac{1}{\sqrt[3]{x^2}}\left(\dfrac{3 + \ln 2x}{3}\right)$

12. Find $\dfrac{dy}{dx}$ given that $y = \ln(1 + 2x)^2$

Solution

$y = \ln(1 + 2x)^2$ [Also $y = \log_e(1 + 2x)^2$]

Let $u = (1 + 2x)^2$

Hence, $y = \ln u$

$$\frac{du}{dx} = 2(1 + 2x) \times \frac{d(2x)}{dx}$$

$$= 2(1 + 2x) \times 2x$$

$$\frac{du}{dx} = 4(1 + 2x)$$

$$\frac{dy}{du} = \frac{1}{u}$$

$$\frac{dy}{dx} = \frac{1}{u} \times 4(1 + 2x)$$

$$= \frac{4(1+2x)}{u}$$

$$= \frac{4(1+2x)}{(1+2x)^2}$$

$$= \frac{4}{1+2x}$$

13. Differentiate with respect to x: $5x\ln(3x^2 - 2)$

<u>Solution</u>

$y = 5x\ln(3x^2 - 2)$

By using product rule:

$u = 5x$

$v = \ln(3x^2 - 2)$

$$\frac{du}{dx} = 5$$

$$\frac{dv}{dx} = \frac{6x}{3x^2 - 2}$$

Hence, $\frac{dy}{dx} = u\frac{dv}{dx} + v\frac{du}{dx}$

$$= 5x\left(\frac{6x}{3x^2 - 2}\right) + \ln(3x^2 - 2)(5)$$

$$\frac{dy}{dx} = \frac{30x^2}{3x^2 - 2} + 5\ln(3x^2 - 2)$$

14. Find the derivative of $\frac{\ln x}{x}$

<u>Solution</u>

$$y = \frac{\ln x}{x}$$

We are going to use quotient rule as follows:

$u = \ln x$

$v = x$

$$\frac{du}{dx} = \frac{1}{x}$$

$$\frac{dv}{dx} = 1$$

Hence, $\dfrac{dy}{dx} = \dfrac{v\dfrac{du}{dx} - u\dfrac{dv}{dx}}{v^2}$

$$= \frac{x\left(\frac{1}{x}\right) - \ln x\ (1)}{x^2}$$

$$\frac{dy}{dx} = \frac{1 - \ln x}{x^2}$$

Exercise 12

1. Find the derivative of $\log_a 7x$.

2. Find the derivative of $\log_a(x^3 + 5)$

3. Differentiate $\log_a(2x^3 - 5)^6$ with respect to x

4. If $y = \log_a x^{\frac{2}{3}}$, find $\dfrac{dy}{dx}$

5. Find $\dfrac{dy}{dx}$ given that $y = \log_a\dfrac{1 + x^2}{1 - x^2}$

6. If $y = \log_5(5x^3 - 1)$, find $\dfrac{dy}{dx}$.

7. Find $\dfrac{dy}{dx}$ if $y = \log_2\dfrac{1}{x^2}$

8. Find the derivative of $\log_e(x^2 + 3)$

9. Find the derivative of $(\log_e x)^3$

10. If $y = \ln\sqrt{1 - 2x^5}$, find $\dfrac{dy}{dx}$.

11. Find $\dfrac{dy}{dx}$ if $y = x^4 \ln x^2$

12. Find $\dfrac{dy}{dx}$ given that $y = \ln(3 - 7x)^4$

13. Differentiate with respect to x: $3x^2 \ln(4x^3 + 1)$

14. Find the derivative of $\dfrac{\ln x^2}{x^2}$

15. Find the derivative of $\log_e(1 - 5x^2)^3$

CHAPTER 13
DERIVATIVE OF EXPONENTIAL FUNCTIONS

If $y = a^x$, then $\dfrac{dy}{dx} = a^x \log_e a$, where 'a' is any number.

If $y = e^x$, then $\dfrac{dy}{dx} = e^x$

Examples

1. Differentiate with respect to x: a^{5x}

Solution

$y = a^{5x}$

Let $u = 5x$

$y = a^u$

$\dfrac{du}{dx} = 5$

$\dfrac{dy}{du} = a^u \log_e a$

Hence, $\dfrac{dy}{dx} = \dfrac{dy}{du} \;\; \text{x} \;\; \dfrac{du}{dx}$

$= a^u \log_e a \;\; \text{x} \;\; 5$

$= 5a^u \log_e a$

$\dfrac{dy}{dx} = 5a^{5x} \log_e a \qquad$ (since $u = 5x$)

2. Find the derivative of $a^{x^2 - 3x + 4}$

Solution

$y = a^{x^2 - 3x + 4}$

Let us solve this example directly without using $u = x^2 - 3x + 4$

$\dfrac{dy}{dx} = \dfrac{d(x^2 - 3x + 4)}{dx} \;\; \text{x} \;\; a^{x^2 - 3x + 4} \;\; \text{x} \log_e a$

$= (2x - 3)(a^{x^2 - 3x + 4} \;\; \text{x} \; \log_e a)$

3. If $y = 3x^2 a^{5x}$, find $\dfrac{dy}{dx}$.

Solution

$y = 3x^2 a^{5x}$

We apply product rule as follows:

$u = 3x^2$

$v = a^{5x}$

$$\frac{du}{dx} = 6x$$

$$\frac{dv}{dx} = 5a^{5x}\log_e a$$

Hence, $\frac{dy}{dx} = u\frac{dv}{dx} + v\frac{du}{dx}$

$$= 3x^2(5a^{5x}\log_e a) + a^{5x}(6x)$$

$$= 15x^2 a^{5x}\log_e a + 6xa^{5x}$$

Factorizing the expression above gives:

$$\frac{dy}{dx} = 3xa^{5x}(5x\log_e a + 2)$$

4. Find the derivative of $\dfrac{e^x + e^{-x}}{e^x - e^{-x}}$

Solution

$$y = \frac{e^x + e^{-x}}{e^x - e^{-x}}$$

$$u = e^x + e^{-x}$$

$$v = e^x - e^{-x}$$

$\dfrac{du}{dx} = e^x - e^{-x}$ (The derivative of $e^{-x} = -1 \times e^{-x} = -e^{-x}$. The value -1 is from the derivative of $-x$)

$$\frac{dv}{dx} = e^x + e^{-x}$$

Hence, $\dfrac{dy}{dx} = \dfrac{v\frac{du}{dx} - u\frac{dv}{dx}}{v^2}$ (Quotient rule)

$$= \frac{(e^x - e^{-x})(e^x - e^{-x}) - (e^x + e^{-x})(e^x + e^{-x})}{(e^x - e^{-x})^2}$$

Expanding the numerator and denominator gives:

$$\frac{dy}{dx} = \frac{(e^x)(e^x) - (e^x)(e^{-x}) - (e^{-x})(e^x) + (e^{-x})(e^{-x}) - [(e^x)(e^x) + (e^x)(e^{-x}) + (e^{-x})(e^x) + (e^{-x})(e^{-x})]}{(e^x - e^{-x})^2}$$

$$= \frac{e^{2x} - 1 - 1 + e^{-2x} - (e^{2x} + 1 + 1 + e^{-2x})}{(e^x - e^{-x})^2}$$ (Note that exponents have been added, and $e^0 =$ 1)

$$= \frac{e^{2x} - 2 + e^{-2x} - e^{2x} - 1 - 1 - e^{-2x})}{(e^x - e^{-x})^2}$$

$\dfrac{dy}{dx} = \dfrac{-4}{(e^x - e^{-x})^2}$ (e^{2x} and e^{-2x} cancel out)

5. If $y = e^{\sqrt{x}}$ find $\dfrac{dy}{dx}$.

Solution

$$y = e^{\sqrt{x}} = e^{x^{\frac{1}{2}}}$$

$$\frac{dy}{dx} = \frac{d(x)^{\frac{1}{2}}}{dx} \times e^{x^{\frac{1}{2}}}$$

$$= \frac{1}{2} x^{-\frac{1}{2}} \times e^{x^{\frac{1}{2}}}$$

$$= \frac{1}{2}\left(\frac{1}{x^{\frac{1}{2}}}\right) \times e^{x^{\frac{1}{2}}}$$

$$= \frac{e^{x^{\frac{1}{2}}}}{2x^{\frac{1}{2}}}$$

$$\frac{dy}{dx} = \frac{e^{\sqrt{x}}}{2\sqrt{x}}$$

6. Find $\frac{dy}{dx}$ if $y = x^{\frac{1}{2}} e^{x^{\frac{1}{2}}}$

Solution

$$y = x^{\frac{1}{2}} e^{x^{\frac{1}{2}}}$$

From product rule:

$$\frac{dy}{dx} = x^{\frac{1}{2}} \frac{d(e^{x^{\frac{1}{2}}})}{dx} + e^{x^{\frac{1}{2}}} \frac{d(x^{\frac{1}{2}})}{dx}$$

$$= x^{\frac{1}{2}}\left(\frac{e^{x^{\frac{1}{2}}}}{2x^{\frac{1}{2}}}\right) + e^{x^{\frac{1}{2}}}\left(\frac{1}{2}x^{-\frac{1}{2}}\right) \quad \text{(Note that the derivative of } e^{x^{\frac{1}{2}}} \text{ is } \frac{e^{x^{\frac{1}{2}}}}{2x^{\frac{1}{2}}} \text{ from example 5)}$$

$$= \frac{e^{x^{\frac{1}{2}}}}{2} + \frac{e^{x^{\frac{1}{2}}}}{2x^{\frac{1}{2}}}$$

$$= \frac{x^{\frac{1}{2}}e^{x^{\frac{1}{2}}} + e^{x^{\frac{1}{2}}}}{2x^{\frac{1}{2}}}$$

$$\frac{dy}{dx} = \frac{e^{x^{\frac{1}{2}}}(x^{\frac{1}{2}} + 1)}{2x^{\frac{1}{2}}}$$

7. Find the derivative of $a^{2x} - a^{-2x}$

Solution

$$y = a^{2x} - a^{-2x}$$

$$\frac{dy}{dx} = \frac{d(2x)}{dx}(a^{2x}\log_e a) - \left[\frac{d(-2x)}{dx}(a^{-2x}\log_e a)\right]$$

$$= 2a^{2x}\log_e a - (-2a^{-2x}\log_e a)$$

$$= 2a^{2x}\log_e a + 2a^{-2x}\log_e a$$

$$= 2\log_e a(a^{2x} + a^{-2x})$$

8. Find the derivative of $e^{2x}\log_e 3x$

Solution

$$y = e^{2x}\log_e 3x$$

By product rule:

$$u = e^{2x}$$

$$v = \log_e 3x \quad (\text{or } v = \ln 3x)$$

$$\frac{du}{dx} = 2e^{2x}$$

$$\frac{dv}{dx} = \frac{3}{3x}$$

$$= \frac{1}{x}$$

$$\frac{dy}{dx} = e^{2x}\left(\frac{1}{x}\right) + \log_e 3x(2e^{2x})$$

$$= \frac{e^{2x}}{x} + 2e^{2x}\log_e 3x$$

$$\frac{dy}{dx} = e^{2x}(\frac{1}{x} + 2\log_e 3x)$$

9. Differentiate with respect to x: $2\sqrt{xe^x}$

Solution

$$y = 2\sqrt{xe^x}$$

Or, $y = 2\sqrt{x} \sqrt{e^x}$

From product rule:

$$u = 2\sqrt{x} = 2x^{\frac{1}{2}}$$

$$v = \sqrt{e^x} = (e^x)^{\frac{1}{2}}$$

$$v = e^{\frac{1}{2}x} \quad \text{(When the exponents are multiplied)}$$

$$\frac{du}{dx} = \frac{1}{2} \times 2x^{-\frac{1}{2}}$$

$$= x^{-\frac{1}{2}}$$

$$\frac{dv}{dx} = \frac{1}{2} e^{\frac{1}{2}x} \quad (\text{Note that } \frac{1}{2} \text{ is from the derivative of } \frac{1}{2}x)$$

$$\frac{dy}{dx} = u\frac{dv}{dx} + v\frac{du}{dx}$$

$$= 2x^{\frac{1}{2}}(\frac{1}{2} e^{\frac{1}{2}x}) + e^{\frac{1}{2}x}(x^{-\frac{1}{2}})$$

$$= (\frac{1}{2} \times 2x^{\frac{1}{2}} \times e^{\frac{1}{2}x}) \times \frac{e^{\frac{1}{2}x}}{x^{\frac{1}{2}}}$$

$$= x^{\frac{1}{2}}e^{\frac{1}{2}x} + \frac{e^{\frac{1}{2}x}}{x^{\frac{1}{2}}}$$

$$= \frac{x\,e^{\frac{1}{2}x} + e^{\frac{1}{2}x}}{x^{\frac{1}{2}}}$$

$$= \frac{e^{\frac{1}{2}x}(x+1)}{x^{\frac{1}{2}}}$$

$$= \frac{e^{\frac{x}{2}}(x+1)}{x^{\frac{1}{2}}}$$

$$\frac{dy}{dx} = \frac{\sqrt{e^x}(x+1)}{\sqrt{x}}$$

10. Find the derivative of $2x\log_{10}x$

Solution

$y = 2x\log_{10}x$ (Note that this is similar to $2x\log_a x$)

$u = 2x$

$v = \log_{10}x$

$\dfrac{du}{dx} = 2$

$\dfrac{dv}{dx} = \dfrac{1}{x}\log_{10}e$ (Recall that the derivative of $\log_a x$ is $\dfrac{1}{x}\log_a e$)

$\dfrac{dy}{dx} = u\dfrac{dv}{dx} + v\dfrac{du}{dx}$

$\quad = 2x\left(\dfrac{1}{x}\log_{10}e\right) + \log_{10}x(2)$

$\quad = 2\log_{10}e + 2\log_{10}x$

$\dfrac{dy}{dx} = 2(\log_{10}e + \log_{10}x)$

Or, $\dfrac{dy}{dx} = 2\left(\dfrac{\log_e e}{\log_e 10} + \log_{10}x\right)$

Note that the rule of change of base in logarithm that has been applied above is given by:

If $\log_a b$ is to be converted to a new base, e, then: $\log_a b = \dfrac{\log_e b}{\log_e a}$

$\dfrac{dy}{dx} = 2\left(\dfrac{1}{\log_e 10} + \log_{10}x\right)$ (Recall that $\log_x x = 1$, or $\log_e e = 1$)

11. Find the derivative of 10^{2x}

Solution

$y = 10^{2x}$ (This is similar to a^{2x})

$\dfrac{dy}{dx} = \dfrac{d(2x)}{dx} \times 10^{2x}\log_e 10$

$= 2 \times 10^{2x}\log_e 10$

$= (10^{2x})2\log_e 10$

$= 10^{2x}\log_e 10^2$ (Recall that $a\log_x y = \log_x y^a$)

$\dfrac{dy}{dx} = 10^{2x}\log_e 100$

12. If $y = e^{\ln x}$ find $\dfrac{dy}{dx}$

Solution

$y = e^{\ln x}$ (Or $y = e^{\log_e x}$ since $\ln x = \log_e x$)

Let, $u = \ln x$

$y = e^u$

$\dfrac{du}{dx} = \dfrac{1}{x}$

$\dfrac{dy}{du} = e^u$

$\dfrac{dy}{dx} = e^u \times \dfrac{1}{x}$

$= \dfrac{e^u}{x}$

$= \dfrac{e^{\ln x}}{x}$ (Since $u = \ln x$)

$= \dfrac{e^{\log_e x}}{x}$ (Since $\ln x = \log_e x$)

$= \dfrac{x}{x}$ (Recall the identity that $m^{\log_m n} = n$)

$\dfrac{dy}{dx} = 1$

This example shows that calculus works, since the derivative of x is 1, and the derivative of $e^{\log_e x} = 1$ because $e^{\log_e x} = x$. Hence, we could have solved this example by stating that:

$\dfrac{d(e^{\ln x})}{dx} = \dfrac{d(x)}{dx} = 1$

13. What is the derivative of $\dfrac{e^{\frac{1}{x}}}{x^2}$

Solution

$y = \dfrac{e^{\frac{1}{x}}}{x^2}$

From quotient rule:

$u = e^{\frac{1}{x}}$

$v = x^2$

$$\frac{du}{dx} = \frac{d\left(\frac{1}{x}\right)}{dx} \; \text{x} \; e^{\frac{1}{x}}$$

$$= \frac{d(x^{-1})}{dx} \; \text{x} \; e^{\frac{1}{x}}$$

$$= -1x^{-2} \; \text{x} \; e^{\frac{1}{x}}$$

$$= \frac{-1}{x^2} \; \text{x} \; e^{\frac{1}{x}}$$

$$\frac{du}{dx} = \frac{-e^{\frac{1}{x}}}{x^2}$$

$$\frac{du}{dx} = 2x$$

$$\frac{dy}{dx} = \frac{v\frac{du}{dx} - u\frac{dv}{dx}}{v^2} \qquad \text{(Quotient rule)}$$

$$= \frac{x^2\left(\frac{-e^{\frac{1}{x}}}{x^2}\right) - e^{\frac{1}{x}}(2x)}{(x^2)^2}$$

$$= \frac{-e^{\frac{1}{x}} - 2x\,e^{\frac{1}{x}}}{x^4} \qquad \text{(Note that } x^2 \text{ has cancelled out)}$$

14. Find the derivative of $e^{-5x} + 5e$

<u>Solution</u>

$$y = e^{-5x} + 5e$$

$$\frac{dy}{dx} = \frac{d(-5x)}{dx} \; \text{x} \; e^{-5x} + \frac{d(5e)}{dx}$$

$$= -5e^{-5x} + 0 \qquad \text{(Note that 5e is a constant and the derivative of a constant is zero)}$$

$$\frac{dy}{dx} = -5e^{-5x}$$

15. If $y = e^{(2x+3)^2}$ find $\frac{dy}{dx}$.

<u>Solution</u>

$$y = e^{(2x+3)^2}$$

Let $u = (2x + 3)^2$

Hence, $y = e^u$

$$\frac{du}{dx} = 2(2x+3)^{2-1} \; \text{x} \; \frac{d(2x)}{dx} \qquad \text{(Using chain rule)}$$

$$= 2(2x+3) \; \text{x} \; 2$$

$$= 4(2x+3)$$

$$\frac{dy}{du} = e^u$$

$$\frac{dy}{dx} = \frac{dy}{du} \times \frac{du}{dx}$$

$$= e^u \times 4(2x + 3)$$

$$= e^{(2x+3)^2} \times 4(2x + 3) \qquad \text{(Note that } u = (2x + 3)^2)$$

$$\frac{dy}{dx} = 4(2x + 3)e^{(2x+3)^2}$$

16. Differentiate with respect to θ, $5^{2\theta}$

Solution

$$y = 5^{2\theta} \qquad \text{(This is like } y = a^{2\theta})$$

$$\frac{dy}{d\theta} = \frac{d(2\theta)}{d\theta} \times 5^{2\theta}\log_e 5$$

$$= 2 \times 5^{2\theta}\log_e 5$$

$$= (5^{2\theta})2\log_e 5$$

$$= 5^{2\theta}\log_e 5^2 \qquad \text{(Note that } 2\log_e 5 = \log_e 5^2)$$

$$\frac{dy}{d\theta} = 5^{2\theta}\log_e 25$$

Exercise 13

1. Differentiate with respect to x: $5a^{2x}$

2. Find the derivative of $a^{5x^3 - x}$

3. If $y = x^4 a^{3x}$, find $\frac{dy}{dx}$.

4. Find the derivative of $\frac{3 + e^{-x}}{e^x}$

5. If $y = 2e^{\sqrt{3x}}$ find $\frac{dy}{dx}$.

6. Find $\frac{dy}{dx}$ if $y = 6x^3 e^{x^2}$

7. Find the derivative of $a^{-5x} + a^{5x}$

8. Find the derivative of $3e^{4x}\log_e 2x^2$

9. Differentiate with respect to x: $3x^2\sqrt{e^{5x}}$

10. Find the derivative of $x^4\log_{10} 7x$

11. Find the derivative of 6^{3x}

12. Find the derivative of $e^{x^2 - 3x}$

13. What is the derivative of $\dfrac{2e^{\frac{3}{x}}}{10x^5}$

14. Find the derivative of $1 - 2e^{-3x}$

113

15. If $y = e^{(x^2+x)^3}$ find $\frac{dy}{dx}$.

16. Differentiate with respect to x, 2^{x^2}

17. Find the derivative of $\dfrac{a^x - a^{-x}}{a^{-x}}$

18. Find the derivative of $e^x \ln 10x^3$

19. Differentiate with respect to x: $e^{2x}\sqrt{5x}$

20. Find the derivative of $2x^5 \log_{10} 5x^4$

CHAPTER 14

LOGARITHMIC DIFFERENTIATION

A function of x could be raised to an exponent which is also a function of x. An example is x^x.

The differentiation of such a function is called logarithmic differentiation.

This type of function is differentiated by first taking the logarithm of both sides of the equation, and then differentiating implicitly before making $\dfrac{dy}{dx}$ the subject of the formula (i.e. solving for $\dfrac{dy}{dx}$). Finally, y (i.e. the function) is substituted into the final answer.

Examples

1. Find the derivative of x^x.

<u>Solution</u>

$$y = x^x$$

This is a function of x raised to an exponent which is also a function of x.

Hence, take the logarithm to base e of both sides as follows:

$$\log_e y = \log_e x^x$$
$$\log_e y = x\log_e x \quad \text{(Note that } \log_m n^b = b\log_m n)$$

Now differentiate y implicitly and differentiate the right hand side using product rule. This gives:

$$\frac{1}{y}\frac{dy}{dx} = x\frac{1}{x} + \log_e x(1) \quad \text{(Use of product rule on the right hand side, with u = } x \text{ and v = } \log_e x)$$

$$\frac{1}{y}\frac{dy}{dx} = 1 + \log_e x$$

Multiply both sides by y to make $\dfrac{dy}{dx}$ the subject of the formula as follows:

$$\frac{dy}{dx} = y(1 + \log_e x)$$
$$\frac{dy}{dx} = x^x(1 + \log_e x) \quad \text{(Since y = } x^x)$$

2. Find the derivative of $x^{\ln x}$.

<u>Solution</u>

$$y = x^{\ln x}$$

Taking the logarithm of both sides gives:

$$\log_e y = \log_e x^{\ln x}$$
$$\log_e y = \ln x\log_e x$$

Differentiate y implicitly and treat the right hand side using product rule as follows:

$$\frac{1}{y}\frac{dy}{dx} = \ln x\left(\frac{1}{x}\right) + \log_e x\left(\frac{1}{x}\right)$$

$$\frac{1}{y}\frac{dy}{dx} = \left(\frac{\ln x}{x}\right) + \left(\frac{\ln x}{x}\right)$$ (Note that $\log_e x$ can also be written as $\ln x$)

$$\frac{1}{y}\frac{dy}{dx} = \frac{2\ln x}{x}$$

Multiply both sides by y. This gives:

$$\frac{dy}{dx} = y\left(\frac{2\ln x}{x}\right)$$

$$\frac{dy}{dx} = x^{\ln x}\left(\frac{2\ln x}{x}\right)$$ (Since $y = x^{\ln x}$)

Or, $\dfrac{dy}{dx} = \dfrac{2}{x}x^{\ln x}\ln x$

3. If $y = \ln(2 + x)^x$ find $\dfrac{dy}{dx}$.

Solution

$y = \ln(2 + x)^x$

In this case we are not going to take the logarithm of both sides since there is already logarithm on the right hand side. Hence, we proceed as follows:

$y = \ln(2 + x)^x$

$y = x\log_e(2 + x)$ [Note that $\ln(2 + x)$ can also be written as $\log_e(2 + x)$]

Using product rule gives:

$$\frac{dy}{dx} = x\left(\frac{1}{2 + x}\right) + \log_e(2 + x)(1)$$

$$\frac{dy}{dx} = \frac{x}{2 + x} + \log_e(2 + x)$$

4. Find $\dfrac{dy}{dx}$ if $y = (x^2 - 3)^{5x}$

Solution

$y = (x^2 - 3)^{5x}$

Taking the natural logarithm of both sides gives:

$\log_e y = \log_e(x^2 - 3)^{5x}$

$\log_e y = 5x\log_e(x^2 - 3)$

Using implicit differentiation for the left hand side and product rule for the right hand side gives:

$$\frac{1}{y}\frac{dy}{dx} = 5x\left(\frac{2x}{x^2 - 3}\right) + \log_e(x^2 - 3)(5)$$

$$\frac{1}{y}\frac{dy}{dx} = \frac{10x^2}{x^2 - 3} + 5\log_e(x^2 - 3)$$

Multiply both sides by y to obtain:

$$\frac{dy}{dx} = y\left[\frac{10x^2}{x^2 - 3} + 5\log_e(x^2 - 3)\right]$$

$$\frac{dy}{dx} = (x^2 - 3)^{5x}\left[\frac{10x^2}{x^2 - 3} + 5\log_e(x^2 - 3)\right] \qquad \text{[Note that } y = (x^2 - 3)^{5x}\text{]}$$

5. If $y = \dfrac{(x^2 - 5)(3x - 1)^2}{x^7(2x^3 - 3)}$

<u>Solution</u>

This is not a function of x raised to an exponent which is also a function of x. This function can be differentiated by the use of product and quotient rule. However, the use of these rules will be a nightmare. Hence, we have to apply logarithmic differentiation. This is done as follows:

$$y = \frac{(x^2 - 5)(3x - 1)^2}{x^7(2x^3 - 3)}$$

Taking the logarithm of both sides gives:

$$\log_e y = \log_e\left[\frac{(x^2 - 5)(3x - 1)^2}{x^7(2x^3 - 3)}\right]$$

In order to continue, we have to recall the theory of logarithm as follows:

1. $\log_b(cd) = \log_b c + \log_b d$

2. $\log_b\left(\dfrac{c}{d}\right) = \log_b c - \log_b d$

3. $\log_b c^d = d\log_b c$

Therefore, we continue the logarithmic differentiation as follows:

$$\log_e y = \log_e\left[\frac{(x^2 - 5)(3x - 1)^2}{x^7(2x^3 - 3)}\right]$$

Applying the theory above gives:

$$\log_e y = \log_e[(x^2 - 5)(3x - 1)^2] - \log_e[x^7(2x^3 - 3)]$$
$$\log_e y = \log_e(x^2 - 5) + \log_e(3x - 1)^2 - [\log_e x^7 + \log_e(2x^3 - 3)]$$
$$\log_e y = \log_e(x^2 - 5) + 2\log_e(3x - 1) - 7\log_e x - \log_e(2x^3 - 3)$$

Differentiating each term accordingly gives:

$$\frac{1}{y}\frac{dy}{dx} = \frac{2x}{x^2 - 5} + \frac{2 \times 3}{3x - 1} - \frac{7}{x} - \frac{6x}{2x^3 - 3}$$

Note that each function of x is differentiated first, and the value obtained is then divided by the original function. Hence, the expression above simplifies to gives:

$$\frac{1}{y}\frac{dy}{dx} = \frac{2x}{x^2 - 5} + \frac{6}{3x - 1} - \frac{7}{x} - \frac{6x}{2x^3 - 3}$$

Multiply both sides by y to obtain $\dfrac{dy}{dx}$ as follows:

$$\frac{dy}{dx} = y\left(\frac{2x}{x^2 - 5} + \frac{6}{3x - 1} - \frac{7}{x} - \frac{6x}{2x^3 - 3}\right)$$

Now replace y with its original value from the question as follows:

$$\frac{dy}{dx} = \left[\frac{(x^2 - 5)(3x - 1)^2}{x^7(2x^3 - 3)}\right]\left(\frac{2x}{x^2 - 5} + \frac{6}{3x - 1} - \frac{7}{x} - \frac{6x}{2x^3 - 3}\right)$$

6. If $y = \dfrac{(2x - 1)(x + 4)(6x - 5)}{(3 - x)^2}$

Solution

$$y = \dfrac{(2x - 1)(x + 4)(6x - 5)}{(3 - x)^2}$$

This is a complex function where product rule and quotient rule would be difficult to apply. Hence, we apply logarithmic differentiation as follows:

$$\log_e y = \log_e\left[\dfrac{(2x - 1)(x + 4)(6x - 5)}{(3 - x)^2}\right]$$

Applying the theory of logarithm gives:

$$\log_e y = \log_e[(2x - 1)(x + 4)(6x - 5)] - \log_e(3 - x)^2$$

$$\log_e y = \log_e(2x - 1) + \log_e(x + 4) + \log_e(6x - 5) - 2\log_e(3 - x)$$

Differentiating accordingly gives:

$$\frac{1}{y}\frac{dy}{dx} = \frac{2}{2x - 1} + \frac{1}{x + 4} + \frac{6}{6x - 5} - \frac{2(-1)}{3 - x}$$

$$\frac{1}{y}\frac{dy}{dx} = \frac{2}{2x - 1} + \frac{1}{x + 4} + \frac{6}{6x - 5} + \frac{2}{3 - x}$$

Multiply both sides by y. This gives:

$$\frac{dy}{dx} = y\left(\frac{2}{2x - 1} + \frac{1}{x + 4} + \frac{6}{6x - 5} + \frac{2}{3 - x}\right)$$

Finally replace y with its original expression as follows:

$$\frac{dy}{dx} = \left[\frac{(2x - 1)(x + 4)(6x - 5)}{(3 - x)^2}\right]\left(\frac{2}{2x - 1} + \frac{1}{x + 4} + \frac{6}{6x - 5} + \frac{2}{3 - x}\right)$$

7. Find the derivative of x^{x^2}

Solution

$$y = x^{x^2}$$

$$\log_e y = \log_e x^{x^2}$$

$$\log_e y = x^2\log_e x$$

$$\frac{1}{y}\frac{dy}{dx} = x^2\left(\frac{1}{x}\right) + \log_e x(2x)$$

$$\frac{1}{y}\frac{dy}{dx} = x + 2x\log_e x$$

$$\frac{1}{y}\frac{dy}{dx} = x(1 + 2\log_e x)$$

$$\frac{dy}{dx} = yx(1 + 2\log_e x)$$

$$= x^{x^2}x(1 + 2\log_e x) \qquad \text{(y has been replaced with } x^{x^2})$$

$$\frac{dy}{dx} = x^{x^2+1}(1 + 2\log_e x)$$

Note that the exponents of x^{x^2} and x were added to obtain x^{x^2+1} since x is also x^1

8. If $y = e^{e^x}$ find $\dfrac{dy}{dx}$.

Solution

$\quad y = e^{e^x}$

$\quad \log_e y = \log_e e^{e^x}$

$\quad \log_e y = e^x \log_e e$

$\quad \log_e y = e^x \qquad$ (Recall that logarithm of a number to the same base is 1. Hence, $\log_e e = 1$)

$\quad \dfrac{1}{y}\dfrac{dy}{dx} = e^x \qquad$ (Note that the derivative of e^x is e^x)

$\quad \dfrac{dy}{dx} = y e^x$

$\qquad = e^{e^x} e^x$

$\quad \dfrac{dy}{dx} = e^{e^x + x} \qquad$ (Their exponents have been added)

9. Find the derivative of 5^{2^x}

Solution

$\quad y = 5^{2^x}$

$\quad \log_e y = \log_e 5^{2^x}$

$\quad \log_e y = 2^x \log_e 5$

Differentiate the left hand side implicitly. On the right hand side, differentiate 2^x and take $\log_e 5$ as a constant. This gives:

$\quad \dfrac{1}{y}\dfrac{dy}{dx} = 2^x \log_e 2 \log_e 5$

Note that 2^x is treated in a similar way as a^x, and recall that the derivative of a^x is $a^x \log_e a$. Hence the derivative of 2^x is $2^x \log_e 2$, and $\log_e 5$ multiplies it since it is a constant. Therefore:

$\quad \dfrac{dy}{dx} = y(2^x \log_e 2 \log_e 5)$

$\qquad = 5^{2^x}(2^x \log_e 2 \log_e 5)$

$\quad \dfrac{dy}{dx} = 5^{2^x}(2^x \log_e 2)(\log_e 5)$

10. If $y = x^{\ln(2x^2 - 5)}$, find $\dfrac{dy}{dx}$.

Solution

$\quad y = x^{\ln(2x^2 - 5)}$

$\quad \log_e y = \log_e x^{\ln(2x^2 - 5)}$

$\quad \log_e y = \ln(2x^2 - 5)\log_e x$

$\quad \dfrac{1}{y}\dfrac{dy}{dx} = \ln(2x^2 - 5)\left(\dfrac{1}{x}\right) + \log_e x\left(\dfrac{4x}{2x^2 - 5}\right) \qquad$ (Use of product rule)

119

$$\frac{1}{y}\frac{dy}{dx} = \frac{1}{x}\ln(2x^2 - 5) + \left(\frac{4x}{2x^2-5}\right)\log_e x$$

$$\frac{dy}{dx} = y\left[\frac{1}{x}\ln(2x^2 - 5) + \left(\frac{4x}{2x^2-5}\right)\log_e x\right]$$

$$\frac{dy}{dx} = x^{\ln(2x^2-5)}\left[\frac{\ln(2x^2-5)}{x} + \left(\frac{4x}{2x^2-5}\right)\log_e x\right]$$

Exercise 14

1. Find the derivative of x^{2x}

2. Find the derivative of $3x^{\ln 2x}$.

3. If $y = \ln(1 - 4x^2)^x$ find $\frac{dy}{dx}$.

4. Find $\frac{dy}{dx}$ if $y = (6x^2 - 5)^{3x}$

5. If $y = \frac{(2x^3 + 1)(x-2)^3}{3x^2(x^3-1)^2}$

6. If $y = \frac{(2x - 1)(x^2 - 2)}{(1-x)(x-3)^2}$

7. Find the derivative of $5x^{3x^4}$

8. If $y = e^{e^{3x}}$ find $\frac{dy}{dx}$.

9. Find the derivative of 2^{e^x}

10. If $y = x^{\ln(x^2 - 4x)}$, find $\frac{dy}{dx}$.

11. If $y = \ln(10 + 2x^2)^x$ find $\frac{dy}{dx}$.

12. Find $\frac{dy}{dx}$ if $y = (3x^3 - 8)^{2x}$

13. If $y = \frac{(x + 3)^2(x-3)^2}{(x^3-1)}$

14. If $y = 5x^{(\ln x)^2}$, find $\frac{dy}{dx}$.

15. Find the derivative of $10x^{3x^2}$

CHAPTER 15
DERIVATIVE OF ONE FUNCTION WITH RESPECT TO ANOTHER

We can differentiate one function with respect to another as illustrated by the examples shown below.

Examples

1. Differentiate x^{12} with respect to x^7.

Solution

Let $u = x^{12}$

And $v = x^7$

$$\frac{du}{dx} = 12x^{11}$$

$$\frac{dv}{dx} = 7x^6$$

Differentiating x^{12} with respect to x^7 means differentiating u (i.e. x^{12}) with respect to v (i.e. x^7).

This means $\frac{du}{dv}$. This follows the parametric equation rule given by:

$$\frac{du}{dv} = \frac{\frac{du}{dx}}{\frac{dv}{dx}}$$

$$= \frac{12x^{11}}{7x^6}$$

$$= \frac{12}{7}x^5$$

2. Differentiate $2e^x$ with respect to $\ln 2x$.

Solution

The differentiation of $2e^x$ with respect to $\ln 2x$ simply means:

$$\frac{\text{The derivative of } 2e^x}{\text{The derivative of } \ln 2x}$$

Hence, $\frac{d(2e^x)}{dx} = 2e^x$

And, $\frac{d(\ln 2x)}{dx} = \frac{2}{2x} = \frac{1}{x}$

Therefore, the differentiation of $2e^x$ with respect to $\ln 2x$ is given by:

$$\frac{2e^x}{\frac{1}{x}}$$

$$= 2xe^x$$

3. Differntiate $\sin 5x$ with respect to $\cos x$.

Solution

Hence, $\dfrac{d(\sin 5x)}{dx} = 5\cos 5x$

And, $\dfrac{d(\cos x)}{dx} = -\sin x$

Therefore, the differentiation of sin5x with respect to cosx is given by:

$$\dfrac{\text{The derivative of } \sin 5x}{\text{The derivative of } \cos x}$$

$$= \dfrac{5\cos 5x}{-\sin x}$$

$$= -\dfrac{5\cos 5x}{\sin x}$$

4. Find the derivative of $2x^2 - 5$ with respect to $4x - 1$

Solution

This is obtained as follows:

$$\dfrac{\frac{d(2x^2 - 5)}{dx}}{\frac{d(4x - 1)}{dx}}$$

$$= \dfrac{4x}{4}$$

$$= x$$

Exercise 15

1. Differentiate x^3 with respect to x^5.
2. Differentiate e^{5x} with respect to $\ln x^2$
3. Differentiate $\cos x^2$ with respect to $\sin x$.
4. Find the derivative of $4x^3 - 5x^2$ with respect to $(x - 1)^2$
5. Differentiate $7x^4$ with respect to $3x^2$.
6. Differentiate $2a^x$ with respect to e^x.
7. Differentiate $\tan^2 x$ with respect to $\cos 5x$.
8. Find the derivative of $\ln(x - 1)$ with respect to e^{x-1}
9. Differentiate $\log_e x^5$ with respect to e^{5x}.
10. Differentiate $\ln(e^x - e^{-x})$ with respect to $\ln x^x$

CHAPTER 16
HIGHER DERIVATIVES (SUCCESSIVE DIFFERENTIATION)

If f(x) is differentiated it gives the first derivative denoted by f'(x) or $\frac{dy}{dx}$. If we differentiate f'(x),

it gives the second derivative denoted by f''(x) or $\frac{d^2y}{dx^2}$ (read as, dee two y dee x squared). Other

higher derivatives such as $\frac{d^3y}{dx^3}, \frac{d^4y}{dx^4}$ etc can also be obtained depending on the function.

Examples

1. Find the first, second and third derivatives of $2x^5 - 3x^4 + 5x^2 - 6$

<u>Solution</u>

$$y = 2x^5 - 3x^4 + 5x^2 - 6$$

$$\frac{dy}{dx} = 10x^4 - 12x^3 + 10x$$

$\frac{d^2y}{dx^2}$ is obtained by differentiating $\frac{dy}{dx}$, which means to differentiate $10x^4 - 12x^3 + 10x$.

Hence, $\frac{d^2y}{dx^2} = 40x^3 - 36x^2 + 10$

$\frac{d^3y}{dx^3}$ is obtained by differentiating $40x^3 - 36x^2 + 10$ as follows:

$$\frac{d^3y}{dx^3} = 120x^2 - 72x$$

In summary, the first derivative $\left(\frac{dy}{dx}\right)$ is $10x^4 - 12x^3 + 10x$, the second derivative $\left(\frac{d^2y}{dx^2}\right)$ is $40x^3 -$

$36x^2 + 10$, while the third derivative $\left(\frac{d^3y}{dx^3}\right)$ is $120x^2 - 72x$.

2. If $y = \ln x^2$, find $\frac{d^2y}{dx^2}$.

<u>Solution</u>

$$y = \ln x^2$$

$$\frac{dy}{dx} = \frac{2x}{x^2}$$

$$= \frac{2}{x}$$

$$\frac{d^2y}{dx^2} = \frac{d\left(\frac{2}{x}\right)}{dx}$$

$$= \frac{d(2x^{-1})}{dx}$$

$$= -2x^{-2}$$

$$\frac{d^2y}{dx^2} = \frac{-2}{x^2}$$

3. Find $\frac{d^3y}{dx^3}$ given that $y = e^{x^3}$

Solution

$y = e^{x^3}$

$\frac{dy}{dx} = 3x^2 e^{x^3}$ (Note that $3x^2$ is from the derivative of x^3)

We now use product rule to obtain $\frac{d^2y}{dx^2}$ as follows:

$$\frac{d^2y}{dx^2} = 3x^2 \left[\frac{d(e^{x^3})}{dx}\right] + e^{x^3}\left[\frac{d(3x^2)}{dx}\right]$$

$$= 3x^2 (3x^2 e^{x^3}) + e^{x^3}(6x)$$

$$= 9x^4 e^{x^3} + 6x e^{x^3}$$

$$\frac{d^2y}{dx^2} = e^{x^3}(9x^4 + 6x)$$

Finally, let us also use product rule to obtain $\frac{d^3y}{dx^3}$ as follows:

$$\frac{d^3y}{dx^3} = e^{x^3}\left[\frac{d(9x^4 + 6x)}{dx}\right] + (9x^4 + 6x)\left[\frac{d(e^{x^3})}{dx}\right]$$

$$= e^{x^3}(36x^3 + 6) + (9x^4 + 6x)(3x^2 e^{x^3})$$

$$= 36x^3 e^{x^3} + 6e^{x^3} + 27x^6 e^{x^3} + 18x^3 e^{x^3}$$

$$= 27x^6 e^{x^3} + 36x^3 e^{x^3} + 18x^3 e^{x^3} + 6e^{x^3}$$

$$= 27x^6 e^{x^3} + 54x^3 e^{x^3} + 6e^{x^3}$$

$$\frac{d^3y}{dx^3} = 3e^{x^3}(9x^6 + 18x^3 + 2)$$

4. If $y = \sin 2x^3$, find the third derivative of y.

Solution

$y = \sin 2x^3$

$\frac{dy}{dx} = \cos 2x^3 \frac{d(2x^3)}{dx}$ (Recall that the derivative of $\sin x$ is $\cos x$)

$= \cos 2x^3 (6x^2)$

$\frac{dy}{dx} = 6x^2 \cos 2x^3$

We now use product rule to obtain $\frac{d^2y}{dx^2}$ as follows:

$$\frac{d^2y}{dx^2} = 6x^2\left[\frac{d(\cos 2x^3)}{dx}\right] + \cos 2x^3\left[\frac{d(6x^2)}{dx}\right]$$

$$= 6x^2\left[-\sin 2x^3 \frac{d(2x^3)}{dx}\right] + \cos 2x^3(12x)$$

$$= 6x^2\left[-\sin 2x^3\,(6x^2)\right] + 12x\cos 2x^3$$

$$\frac{d^2y}{dx^2} = -36x^4\sin 2x^3 + 12x\cos 2x^3$$

Again, the use of product rule gives us $\frac{d^3y}{dx^3}$ as follows:

$$\frac{d^3y}{dx^3} = -36x^4\left[\frac{d\left(\sin 2x^3\right)}{dx}\right] + \sin 2x^3\left[\frac{d\left(-36x^4\right)}{dx}\right] + 12x\left[\frac{d\left(\cos 2x^3\right)}{dx}\right] + \cos 2x^3\left[\frac{d(12x)}{dx}\right]$$

Note that the derivative of $\sin 2x^3$ is $6x^2\cos 2x^3$ as obtained from $\frac{dy}{dx}$. Similarly, the differentiation of $\cos 2x^3$ will give us $-6x^2\sin 2x^3$ (since the derivative of $\sin x$ and $\cos x$ differs only by sign and the interchanging of sin with cos or cos with sin).

We now replace the derivative of $\sin 2x^3$ and $\cos 2x^3$ with $6x^2\cos 2x^3$ and $-6x^2\sin 2x^3$ respectively in the expression above. This gives:

$$\frac{d^3y}{dx^3} = -36x^4\,(6x^2\cos 2x^3) + \sin 2x^3(-144x^3) + 12x\,(-6x^2\sin 2x^3) + \cos 2x^3(12)$$

$$= -216x^6\cos 2x^3 - 144x^3\sin 2x^3 - 72x^3\sin 2x^3 + 12\cos 2x^3$$

$$\frac{d^3y}{dx^3} = -216x^6\cos 2x^3 - 216x^3\sin 2x^3 + 12\cos 2x^3$$

Or, $\frac{d^3y}{dx^3} = -12(18x^6\cos 2x^3 + 18x^3\sin 2x^3 - \cos 2x^3)$

5. Find $\frac{d^2y}{dx^2}$ given that $y = a^{x^2} - 5$

Solution

$$y = a^{x^2} - 5$$

$$\frac{dy}{dx} = a^{x^2}\left[\frac{d(x^2)}{dx}\right]\log_e a$$

$$= a^{x^2}(2x)\log_e a$$

$$\frac{dy}{dx} = 2x\,a^{x^2}\log_e a$$

We now use product rule to find $\frac{d^2y}{dx^2}$ as follows:

$$\frac{d^2y}{dx^2} = 2x\left[\frac{d\left(a^{x^2}\log_e a\right)}{dx}\right] + a^{x^2}\log_e a\left[\frac{d(2x)}{dx}\right]$$

$$= 2x\left[\frac{\log_e a\,d\left(a^{x^2}\right)}{dx}\right] + a^{x^2}\log_e a(2)$$

$$= 2x[\log_e a(2x a^{x^2}\log_e a)] + 2a^{x^2}\log_e a$$

$$= 4x^2 a^{x^2}(\log_e a)^2 + 2a^{x^2}\log_e a$$

$$\frac{d^2y}{dx^2} = 2a^{x^2}\log_e a(2x^2\log_e a + 1)$$

6. Find the second derivative of $y = \dfrac{\cos x}{x^3}$

Solution

$$y = \frac{\cos x}{x^3}$$

We use quotient rule as follows:

$$\frac{dy}{dx} = \frac{x^3\frac{d(\cos x)}{dx} - \cos x\frac{d(x^3)}{dx}}{(x^3)^2}$$

$$= \frac{x^3(-\sin x) - \cos x(3x^2)}{x^6}$$

$$\frac{dy}{dx} = \frac{-x^3\sin x - 3x^2\cos x}{x^6}$$

Divide each part by x^6 to separate into fractions as follows:

$$\frac{dy}{dx} = \frac{-x^3\sin x}{x^6} - \frac{3x^2\cos x}{x^6}$$

$$\frac{dy}{dx} = \frac{-\sin x}{x^3} - \frac{3\cos x}{x^4}$$

We now apply quotient rule again as follows:

$$\frac{d^2y}{dx^2} = \frac{x^3\frac{d(-\sin x)}{dx} - (-\sin x)\frac{d(x^3)}{dx}}{(x^3)^2} - \left[\frac{x^4\frac{d(3\cos x)}{dx} - 3\cos x\frac{d(x^4)}{dx}}{(x^4)^2}\right]$$

$$= \frac{x^3(-\cos x) + \sin x\,(3x^2)}{x^6} - \left[\frac{x^4(-3\sin x) - 3\cos x\,(4x^3)}{x^8}\right]$$

$$= \frac{-x^3\cos x + 3x^2\sin x}{x^6} + \frac{3x^4\sin x + 12x^3\cos x}{x^8}$$

Separating into fractions gives:

$$\frac{d^2y}{dx^2} = \frac{-x^3\cos x}{x^6} + \frac{3x^2\sin x}{x^6} + \frac{3x^4\sin x}{x^8} + \frac{12x^3\cos x}{x^8}$$

$$= \frac{-\cos x}{x^3} + \frac{3\sin x}{x^4} + \frac{3\sin x}{x^4} + \frac{12\cos x}{x^5}$$

Combining the fractions again by using x^5 as LCM gives:

$$\frac{d^2y}{dx^2} = \frac{-x^2\cos x + 3x\sin x + 3x\sin x + 12\cos x}{x^5}$$

$$\frac{d^2y}{dx^2} = \frac{6x\sin x + 12\cos x - x^2\cos x}{x^5}$$

7. Find the third derivative of $y = \log_e(1 + 2x)^2$

Solution

$$y = \log_e(1 + 2x)^2$$

$$\frac{dy}{dx} = \frac{2(1+2x)^{2-1} \times \frac{d(2x)}{dx}}{(1+2x)^2}$$ [Note the use of chain rule in finding the derivative of $(1+2x)^2$]

$$= \frac{2(1+2x) \times 2}{(1+2x)^2}$$

$$= \frac{4(1+2x)}{(1+2x)^2}$$

$$= \frac{4}{1+2x} \qquad (1+2x \text{ cancels out})$$

$$\frac{dy}{dx} = 4(1+2x)^{-1}$$

Using chain rule gives $\frac{d^2y}{dx^2}$ as follows:

$$\frac{d^2y}{dx^2} = -1 \times 4(1+2x)^{-1-1} \times \frac{d(2x)}{dx}$$

$$= -4(1+2x)^{-2} \times 2$$

$$= -8(1+2x)^{-2}$$

Use of chain rule again gives $\frac{d^3y}{dx^3}$ as follows:

$$\frac{d^3y}{dx^3} = -2 \times -8(1+2x)^{-2-1} \times \frac{d(2x)}{dx}$$

$$= 16(1+2x)^{-3} \times 2$$

$$= 32(1+2x)^{-3}$$

$$\frac{d^3y}{dx^3} = \frac{32}{(1+2x)^3}$$

8. If y = sec2x, find $\frac{d^3y}{dx^3}$.

Solution

$$y = \sec 2x$$

$$\frac{dy}{dx} = 2\sec 2x \tan 2x$$

We now use product rule to obtain $\frac{d^2y}{dx^2}$ as follows:

$$\frac{d^2y}{dx^2} = 2\sec 2x\frac{d(\tan 2x)}{dx} + \tan 2x\frac{d(2\sec 2x)}{dx}$$

$$= 2\sec 2x(2\sec^2 2x) + \tan 2x(2 \times 2\,\sec 2x\tan 2x)$$

$$\frac{d^2y}{dx^2} = 4\sec^3 2x + 4\sec 2x\tan^2 2x$$

We now use chain rule for $4\sec^3 2x$ and product rule for $4\sec 2x\tan^2 2x$ to obtain $\frac{d^3y}{dx^3}$ as follows:

$$\frac{d^3y}{dx^3} = 3 \times 4\sec^2 2x\frac{d(\sec 2x)}{dx} + 4\sec 2x\frac{d(\tan^2 2x)}{dx} + \tan^2 2x\frac{d(4\sec 2x)}{dx}$$

$$= 12\sec^2 2x(2\sec 2x\tan 2x) + 4\sec 2x\left[2\tan 2x\frac{d(\tan 2x)}{dx}\right] + \tan^2 2x[4(2\sec 2x\tan 2x)]$$

127

$$= 24\sec^3 2x\tan 2x + 8\sec 2x\tan 2x(2\sec^2 2x) + 8\sec 2x\tan^3 2x$$

$$= 24\sec^3 2x\tan 2x + 16\sec^3 2x\tan 2x + 8\sec 2x\tan^3 2x$$

$$\frac{d^3y}{dx^3} = 40\sec^3 2x\tan 2x + 8\sec 2x\tan^3 2x$$

9. Given that y = 2cos5x + 3sin5x, prove that $\frac{d^2y}{dx^2} + 25y = 0$

Solution

   y = 2cos5x + 3sin5x

$$\frac{dy}{dx} = 2 \times 5(-\sin 5x) + 3 \times 5(\cos 5x)$$

$$= -10\sin 5x + 15\cos 5x$$

$$\frac{d^2y}{dx^2} = -10 \times 5(\cos 5x) + 15 \times 5(-\sin 5x)$$

$$= -50\cos 5x - 75\sin 5x$$

We now obtain 25y as follows:

   y = 2cos5x + 3sin5x

25y = 25(2cos5x + 3sin5x)

25y = 50cos5x + 75sin5x

We now simplify $\frac{d^2y}{dx^2} + 25y$ as follows:

$$\frac{d^2y}{dx^2} + 25y = -50\cos 5x - 75\sin 5x + 50\cos 5x + 75\sin 5x$$

$$= 0 \quad \text{(Equal terms with opposite signs cancel out)}$$

Therefore, $\frac{d^2y}{dx^2} + 25y = 0$ (As proven above)

10. If $y = x + \sqrt{4 + x^2}$, show that $(4 + x^2)\frac{d^2y}{dx^2} + x\frac{dy}{dx} - y = 0$

Solution

$$y = x + \sqrt{4 + x^2}$$

$$y = x + (4 + x^2)^{\frac{1}{2}}$$

$$\frac{dy}{dx} = 1 + \frac{1}{2}(4 + x^2)^{\frac{1}{2} - 1} \times \frac{d(x^2)}{dx}$$

$$= 1 + \frac{1}{2}(4 + x^2)^{-\frac{1}{2}} \times 2x$$

$$\frac{dy}{dx} = 1 + x(4 + x^2)^{-\frac{1}{2}}$$

Using product rule, we obtain $\frac{d^2y}{dx^2}$ as follows:

$$\frac{d^2y}{dx^2} = 0 + x\left[\frac{d(4+x^2)^{-\frac{1}{2}}}{dx}\right] + (4+x^2)^{-\frac{1}{2}}\frac{d(x)}{dx}$$

$$= x\left[-\frac{1}{2}(4+x^2)^{-\frac{1}{2}-1}\frac{d(x^2)}{dx}\right] + (4+x^2)^{-\frac{1}{2}}(1)$$

$$= x\left[-\frac{1}{2}(4+x^2)^{-\frac{3}{2}} \times 2x\right] + (4+x^2)^{-\frac{1}{2}}$$

$$= -x^2(4+x^2)^{-\frac{3}{2}} + (4+x^2)^{-\frac{1}{2}}$$

$$\frac{d^2y}{dx^2} = \frac{-x^2}{(4+x^2)^{\frac{3}{2}}} + \frac{1}{(4+x^2)^{\frac{1}{2}}}$$

Let us now simplify $(4+x^2)\frac{d^2y}{dx^2} + x\frac{dy}{dx} - y$ as follows:

$$(4+x^2)\left[\frac{-x^2}{(4+x^2)^{\frac{3}{2}}} + \frac{1}{(4+x^2)^{\frac{1}{2}}}\right] + x[1 + x(4+x^2)^{-\frac{1}{2}}] - [x + (4+x^2)^{\frac{1}{2}}]$$

Expanding the brackets gives:

$$-x^2(4+x^2)^{1-\frac{3}{2}} + (4+x^2)^{1-\frac{1}{2}} + x + x^2(4+x^2)^{-\frac{1}{2}} - x - (4+x^2)^{\frac{1}{2}}$$

$$= -x^2(4+x^2)^{-\frac{1}{2}} + (4+x^2)^{\frac{1}{2}} + x + x^2(4+x^2)^{-\frac{1}{2}} - x - (4+x^2)^{\frac{1}{2}}$$

$$= 0 \quad \text{(Note that equal terms with opposite signs cancel out one another to give zero)}$$

Therefore, $(4+x^2)\frac{d^2y}{dx^2} + x\frac{dy}{dx} - y = 0$ (As proven above)

Exercise 16

1. Find the first, second and third derivatives of $x^4 - 2x^3 - 7x^2 + 1$

2. If $y = \ln 2x^3$, find $\frac{d^2y}{dx^2}$.

3. Find $\frac{d^2y}{dx^2}$ given that $y = e^{5x^4}$

4. If $y = \sin^2 x^5$, find the second derivative of y.

5. Find $\frac{d^3y}{dx^3}$ given that $y = \ln(4x - 10)$

6. Find the second derivative of $y = \frac{\sin x}{x}$

7. Find the third derivative of $y = (\ln x)^2$

8. If $y = \cot x$, find $\frac{d^3y}{dx^3}$.

9. Given that $y = \sin^2 x + \cos 2x$, find $\frac{d^2y}{dx^2} - y$

10. If $y = 2e^{2x} + 5e^{-x}$ evaluate $\dfrac{d^2y}{dx^2} - \dfrac{dy}{dx} - 2y$

11. Given that $y = \cos^2x + 2\sin x^2$ find $\dfrac{d^2y}{dx^2} - \dfrac{dy}{dx}$

12. Find the second derivative of $y = \dfrac{\tan x}{3x}$

13. If $y = e^{-x}\cos 2x$, find the second derivative of y.

14. If $y = 2\cos 3x + 5\sin 3x$, evaluate $\dfrac{d^2y}{dx^2} + 9y$

15. Find the second derivative of $y = \dfrac{\sin 3x}{3x}$

CHAPTER 17
MISCELLANEOUS PROBLEMS ON DIFFERENTIAL CALCULUS

This chapter covers worked examples on general problems involving combination of topics treated in the various previous chapters. More challenging problems would also be covered here.

Examples

1. If $y = \dfrac{1}{(2x^3 - 5)^4}$ find $\dfrac{dy}{dx}$.

Solution

$$y = \frac{1}{(2x^3 - 5)^4}$$

This can be written as:

$$y = (2x^3 - 5)^{-4}$$

We now use chain rule to differentiate it as follows:

$$\frac{dy}{dx} = -4(2x^3 - 5)^{-4-1} \times \frac{d(2x^3 - 5)}{dx}$$

$$= -4(2x^3 - 5)^{-5} \times 6x^2$$

$$= -24x^2(2x^3 - 5)^{-5}$$

$$\frac{dy}{dx} = \frac{-24x^2}{(2x^3 - 5)^5}$$

2. Given that $f(x) = 5x^4 - 3x^3 - 7x^2 + 9$, find $f'(2)$

Solution

$$f(x) = 5x^4 - 3x^3 - 7x^2 + 9$$

We differentiate $f(x)$ to obtain $f'(x)$ as follows:

$$f'(x) = 20x^3 - 9x^2 - 14x$$

In order to find $f'(2)$, we simply substitute 2 for x in $f'(x)$ as follows:

$$f'(x) = 20x^3 - 9x^2 - 14x$$

$$f'(2) = 20(2^3) - 9(2^2) - 14(2)$$

$$= 20(8) - 9(4) - 28$$

$$= 160 - 36 - 28$$

$$f'(2) = 96$$

3. A function $f(x)$ is given by $f(x) = \dfrac{\sqrt{4 + 3x^2}}{x^3}$. Find:

(a) the derivative of $f(x)$

(b) the gradient of $f(x)$ at the point $(2, \dfrac{1}{2})$

Solution

(a) $f(x) = \dfrac{\sqrt{4 + 3x^2}}{x^3}$

We apply quotient rule to differentiate f(x) as follows:

$$f'(x) = \frac{x^3\left[\dfrac{d(\sqrt{4+3x^2})}{dx}\right] - \sqrt{4+3x^2}\left[\dfrac{d(x^3)}{dx}\right]}{(x^3)^2}$$

$$= \frac{x^3\left[\dfrac{d(4+3x^2)^{\frac{1}{2}}}{dx}\right] - (4+3x^2)^{\frac{1}{2}}(3x^2)}{x^6}$$

$$= \frac{x^3\left[\dfrac{1}{2}(4+3x^2)^{\frac{1}{2}-1} \times \dfrac{d(3x^2)}{dx}\right] - 3x^2(4+3x^2)^{\frac{1}{2}}}{x^6}$$

$$= \frac{x^3\left[\dfrac{1}{2}(4+3x^2)^{-\frac{1}{2}} \times 6x\right] - 3x^2(4+3x^2)^{\frac{1}{2}}}{x^6}$$

$$= \frac{x^3\left[3x(4+3x^2)^{-\frac{1}{2}}\right] - 3x^2(4+3x^2)^{\frac{1}{2}}}{x^6}$$

$$= \frac{3x^4(4+3x^2)^{-\frac{1}{2}} - 3x^2(4+3x^2)^{\frac{1}{2}}}{x^6}$$

Let us factorize the expression above by taking $3x^2(4+3x^2)^{-\frac{1}{2}}$ as the common factor. This gives:

$$f'(x) = \frac{3x^2(4+3x^2)^{-\frac{1}{2}}[x^2 - (4+3x^2)]}{x^6}$$

Note that $(4+3x^2)^{\frac{1}{2}} \div (4+3x^2)^{-\frac{1}{2}} = (4+3x^2)^{\frac{1}{2}-(-\frac{1}{2})} = 4+3x^2$ as the exponent becomes 1. Recall that exponents are subtracted during division.

$$= \frac{3x^2(4+3x^2)^{-\frac{1}{2}}(x^2 - 4 - 3x^2)}{x^6}$$

$$= \frac{3x^2(4+3x^2)^{-\frac{1}{2}}(-2x^2 - 4)}{x^6}$$

$$= \frac{-3x^2(4+3x^2)^{-\frac{1}{2}}(2x^2 + 4)}{x^6}$$

$$= \frac{-3(4+3x^2)^{-\frac{1}{2}}(2x^2 + 4)}{x^4} \qquad \text{(Note that } x^2 \text{ has cancelled out of } x^6)$$

$$f'(x) = \frac{-3(2x^2 + 4)}{x^4(4+3x^2)^{\frac{1}{2}}}$$

133

(b) The gradient of f(x) at the point $(2, \frac{1}{2})$ is obtained by substituting 2 for x in f'(x). Note that the point $(2, \frac{1}{2})$ is at $x = 2$ and $y = \frac{1}{2}$. We ignore $y = \frac{1}{2}$ since y is not in the expression for f'(x).

Hence, $f'(x) = \dfrac{-3(2x^2 + 4)}{x^4(4 + 3x^2)^{\frac{1}{2}}}$

At $(2, \frac{1}{2})$, $f'(2) = \dfrac{-3(2(2)^2 + 4)}{(2)^4(4 + 3(2)^2)^{\frac{1}{2}}}$

$= \dfrac{-3(8 + 4)}{16(4 + 12)^{\frac{1}{2}}}$

$= \dfrac{-3(12)}{16(16)^{\frac{1}{2}}}$

$= \dfrac{-36}{16 \text{ x } 4}$ (Note that $(16)^{\frac{1}{2}} = \sqrt{16} = 4$)

$f'(2) = \dfrac{-9}{16}$ (In its lowest term)

Therefore the gradient of f(x) at the point $(2, \frac{1}{2})$ is $\dfrac{-9}{16}$

4. If $x^2y^2 - 3xy + 4xy^3 = 4$, find:
(a) the derivative of the expression
(b) the gradient at $(-1, 2)$
Solution
(a) $x^2y^2 - 3xy + 4xy^3 = 4$
The use of implicit differentiation combined with product rule gives us the derivative as follows:

$x^2\dfrac{d(y^2)}{dx} + y^2\dfrac{d(x^2)}{dx} - \left[3x\dfrac{d(y)}{dx} + y\dfrac{d(3x)}{dx}\right] + 4x\dfrac{d(y^3)}{dx} + y^3(4) = \dfrac{d(4)}{dx}$

$x^2\left(2y\dfrac{dy}{dx}\right) + y^2(2x) - \left[3x\dfrac{dy}{dx} + y(3)\right] + 4x\left(3y^2\dfrac{dy}{dx}\right) + 4y^3 = 0$

$2x^2y\dfrac{dy}{dx} + 2xy^2 - 3x\dfrac{dy}{dx} - 3y + 12xy^2\dfrac{dy}{dx} + 4y^3 = 0$

Collect terms in $\dfrac{dy}{dx}$ on one side of the equation. This gives:

$2x^2y\dfrac{dy}{dx} - 3x\dfrac{dy}{dx} + 12xy^2\dfrac{dy}{dx} = 3y - 2xy^2 - 4y^3$

Factorizing the left hand side gives:

$\dfrac{dy}{dx}(2x^2y - 3x + 12xy^2) = 3y - 2xy^2 - 4y^3$

Divide both sides by the terms in the bracket. This gives:

$\dfrac{dy}{dx} = \dfrac{3y - 2xy^2 - 4y^3}{12x^2y - 3x + 12xy^2}$

(b) At the point (−1, 2) the gradient of the expression is obtained by simply substituting −1 for x and 2 for y in the expression for the derivative. This is done as follows:

$$\frac{dy}{dx} = \frac{3y - 2xy^2 - 4y^3}{12x^2y - 3x + 12xy^2}$$

$$= \frac{3(2) - 2(-1)(2)^2 - 4(2)^3}{12(-1)^2(2) - 3(-1) + 12(-1)(2)^2}$$

$$= \frac{6 + 8 - 32}{24 + 3 - 48}$$

$$= \frac{-18}{-21}$$

$$= \frac{6}{7}$$

5. Find the derivative of $(x + 3y^2)^3 = 7$

Solution

$(x + 3y^2)^3 = 7$

We differentiate implicitly and apply chain rule as follows:

$$3(x + 3y^2)^{3-1} \times \frac{d(x + 3y^2)}{dx} = 0$$

$$3(x + 3y^2)^2\left(1 + 6y\frac{dy}{dx}\right) = 0$$

Dividing both sides by $3(x + 3y^2)^2$ gives:

$$1 + 6y\frac{dy}{dx} = 0 \qquad \left(\text{Note that } \frac{0}{3(x + 3y^2)^2} = 0\right)$$

$$6y\frac{dy}{dx} = -1$$

$$\frac{dy}{dx} = -\frac{1}{6y}$$

6. If y = $(3x^2 - 2x + 5)(2x - 3)$, find $\frac{dy}{dx}$.

Solution

y = $(3x^2 - 2x + 5)(2x - 3)$

We differentiate the expression by applying product rule as follows:

$$\frac{dy}{dx} = (3x^2 - 2x + 5)(2) + (2x - 3)(6x - 2)$$

$$= 6x^2 - 4x + 10 + 12x^2 - 4x - 18x + 6$$

$$\frac{dy}{dx} = 18x^2 - 26x + 16$$

7. Find the derivative of $a^{1 + \tan x}$

Solution

y = $a^{1 + \tan x}$

135

Let u = 1 + tan$x$

Hence, y = a^u

$$\frac{du}{dx} = \sec^2 x$$

$$\frac{dy}{du} = a^u \log_e a$$

Hence, $\frac{dy}{dx} = \frac{dy}{du} \times \frac{du}{dx}$

$= a^u \log_e a \times \sec^2 x$

$$\frac{dy}{dx} = a^{1+\tan x} \sec^2 x \log_e a$$

8. Find the derivative of $\dfrac{\log_e x}{1+\cos x}$

Solution

$$y = \frac{\log_e x}{1+\cos x}$$

We apply product rule as follows:

$$\frac{dy}{dx} = \frac{(1+\cos x)\frac{d(\log_e x)}{dx} - \log_e x \frac{d(1+\cos x)}{dx}}{(1+\cos x)^2}$$

$$= \frac{(1+\cos x)\frac{1}{x} - \log_e x(-\sin x)}{(1+\cos x)^2}$$

$$= \frac{\frac{1+\cos x}{x} + \sin x \log_e x}{(1+\cos x)^2}$$

$$= \frac{\frac{1+\cos x + x \sin x \log_e x}{x}}{(1+\cos x)^2}$$

$$\frac{dy}{dx} = \frac{1+\cos x + x \sin x \log_e x}{x(1+\cos x)^2}$$

9. Differentiate with respect to x: $\ln(\cos x + \sin x)$

Solution

$y = \ln(\cos x + \sin x)$

$$\frac{dy}{dx} = \frac{\frac{d(\cos x + \sin x)}{dx}}{\cos x + \sin x}$$

$$= \frac{-\sin x + \cos x}{\cos x + \sin x}$$

$$\frac{dy}{dx} = \frac{\cos x - \sin x}{\cos x + \sin x}$$

10. If $y = e^{\sin 2x + \cos x}$ find $\frac{dy}{dx}$.

Solution

$$y = e^{\sin 2x + \cos x}$$

$$\frac{dy}{dx} = \frac{d(\sin 2x + \cos x)}{dx} \times e^{\sin 2x + \cos x}$$

$$= (2\cos 2x - \sin x)(e^{\sin 2x + \cos x})$$

11. Given that $y = e^x - e^{-x}$ show that $\frac{d^3y}{dx^3} + \frac{d^2y}{dx^2} + \frac{dy}{dx} + y = 4e^x$

Solution

$$y = e^x - e^{-x}$$

$$\frac{dy}{dx} = e^x - (-e^{-x}) \quad \text{(Note that } \frac{d(e^{-x})}{dx} = \frac{d(-x)}{dx} \times e^{-x} = -1 \times e^{-x} = -e^{-x})$$

$$= e^x + e^{-x}$$

$$\frac{d^2y}{dx^2} = e^x - e^{-x}$$

$$\frac{d^3y}{dx^3} = e^x + e^{-x}$$

Let us now substitute corresponding term into $\frac{d^3y}{dx^3} + \frac{d^2y}{dx^2} + \frac{dy}{dx} + y$ as follows:

$$(e^x + e^{-x}) + (e^x - e^{-x}) + (e^x + e^{-x}) + (e^x - e^{-x})$$

Collecting like terms together gives:

$$e^x + e^x + e^x + e^x + e^{-x} - e^{-x} + e^{-x} - e^{-x}$$

$$= 4e^x \quad \text{(Note that } e^{-x} \text{ cancels out each other)}$$

Hence, $\frac{d^3y}{dx^3} + \frac{d^2y}{dx^2} + \frac{dy}{dx} + y = 4e^x$ (As proven above)

12. Differentiate with respect to x: $\ln\left(\frac{1 - 3x^2}{1 + 3x^2}\right)^{\frac{1}{2}}$

Solution

$$y = \ln\left(\frac{1 - 3x^2}{1 + 3x^2}\right)^{\frac{1}{2}}$$

$$= \log_e\left(\frac{1 - 3x^2}{1 + 3x^2}\right)^{\frac{1}{2}} \quad \text{(Note that "ln" is } \log_e)$$

$$= \log_e\left[\frac{(1 - 3x^2)^{\frac{1}{2}}}{(1 + 3x^2)^{\frac{1}{2}}}\right]$$

$$= \log_e(1 - 3x^2)^{\frac{1}{2}} - \log_e(1 + 3x^2)^{\frac{1}{2}}$$

$$y = \frac{1}{2}\log_e(1 - 3x^2) - \frac{1}{2}\log_e(1 + 3x^2)$$

$$\frac{dy}{dx} = \frac{1}{2}\frac{\frac{d(1-3x^2)}{dx}}{1-3x^2} - \frac{1}{2}\frac{\frac{d(1+3x^2)}{dx}}{1+3x^2}$$

$$= \frac{1}{2}\frac{-6x}{1-3x^2} - \frac{1}{2}\frac{6x}{1+3x^2}$$

$$= \frac{-3x}{1-3x^2} - \frac{3x}{1+3x^2} \qquad \text{(Note that } \frac{1}{2} \text{ reduces } 6x \text{ to } 3x\text{)}$$

$$= \frac{-3x(1+3x^2) - 3x(1-3x^2)}{(1-3x^2)(1+3x^2)}$$

$$= \frac{-3x - 9x^3 - 3x + 9x^3)}{(1-3x^2)(1+3x^2)}$$

$$= \frac{-6x}{1+3x^2 - 3x^2 - 9x^4}$$

$$\frac{dy}{dx} = \frac{-6x}{1-9x^4}$$

13. If $y = \frac{x}{\sqrt{9-x^2}}$ show that: $(9-x^2)\frac{d^2y}{dx^2} = 3x\frac{dy}{dx}$

Solution

$$y = \frac{x}{\sqrt{9-x^2}}$$

Applying quotient rule gives $\frac{dy}{dx}$ as follows:

$$\frac{dy}{dx} = \frac{\sqrt{9-x^2}(1) - x\frac{d(\sqrt{9-x^2})}{dx}}{(\sqrt{9-x^2})^2}$$

$$= \frac{(9-x^2)^{\frac{1}{2}}(1) - x\frac{d[(9-x^2)^{\frac{1}{2}}]}{dx}}{[(9-x^2)^{\frac{1}{2}}]^2}$$

$$= \frac{(9-x^2)^{\frac{1}{2}} - x\left[\frac{1}{2}(9-x^2)^{\frac{1}{2}-1}\right] \times (-2x)}{9-x^2} \qquad \text{(Note that } -2x \text{ is from the derivative of } -x^2\text{)}$$

$$= \frac{(9-x^2)^{\frac{1}{2}} - x\left[-x(9-x^2)^{-\frac{1}{2}}\right]}{9-x^2}$$

$$= \frac{(9-x^2)^{\frac{1}{2}} + x^2(9-x^2)^{-\frac{1}{2}}}{9-x^2}$$

$$= \frac{(9-x^2)^{\frac{1}{2}} + \frac{x^2}{(9-x^2)^{\frac{1}{2}}}}{9-x^2}$$

$$= \frac{\dfrac{9-x^2+x^2}{(9-x^2)^{\frac{1}{2}}}}{9-x^2}$$

$$= \frac{9}{(9-x^2)^{\frac{1}{2}}(9-x^2)}$$

$$= \frac{9}{(9-x^2)^{\frac{3}{2}}}$$

$$\frac{dy}{dx} = 9(9-x^2)^{-\frac{3}{2}}$$

Applying chain rule gives $\dfrac{d^2y}{dx^2}$ as follows:

$$\frac{d^2y}{dx^2} = \frac{-3}{2} \times 9(9-x^2)^{-\frac{3}{2}-1} \times -2x \qquad \text{(Note that } -2x \text{ is from the derivative of } -x^2)$$

$$= \frac{-27}{2} \times -2x(9-x^2)^{-\frac{5}{2}}$$

$$= 27x(9-x^2)^{-\frac{5}{2}}$$

$$\frac{d^2y}{dx^2} = \frac{27x}{(9-x^2)^{\frac{5}{2}}}$$

Let us now show that $(9-x^2)\dfrac{d^2y}{dx^2} = 3x\dfrac{dy}{dx}$

We simplify $(9-x^2)\dfrac{d^2y}{dx^2}$ as follows:

$$(9-x^2)\frac{27x}{(9-x^2)^{\frac{5}{2}}}$$

$$= \frac{27x(9-x^2)}{(9-x^2)^{\frac{5}{2}}}$$

$$= 27x(9-x^2)^{1-\frac{5}{2}}$$

$$= 27x(9-x^2)^{-\frac{3}{2}}$$

$$= \frac{27x}{(9-x^2)^{\frac{3}{2}}}$$

Hence $(9-x^2)\dfrac{d^2y}{dx^2}$ gives us $\dfrac{27x}{(9-x^2)^{\frac{3}{2}}}$

Let us now simplify $3x\dfrac{dy}{dx}$ as follows:

$$3x[9(9-x^2)^{-\frac{3}{2}}] \qquad \text{(Note that } \frac{dy}{dx} = 9(9-x^2)^{-\frac{3}{2}} \text{ as obtained above)}$$

$$= 27x(9-x^2)^{-\frac{3}{2}}$$

$$= \frac{27x}{(9-x^2)^{\frac{3}{2}}}$$

Hence $3x\dfrac{dy}{dx}$ also gives us $\dfrac{27x}{(9-x^2)^{\frac{3}{2}}}$

Therefore, $(9-x^2)\dfrac{d^2y}{dx^2} = 3x\dfrac{dy}{dx}$ as both sides give $\dfrac{27x}{(9-x^2)^{\frac{3}{2}}}$

14. Given that $y = \dfrac{x}{x-1}$, show that: $(x-1)\dfrac{d^2y}{dx^2} + 2\dfrac{dy}{dx} = 0$

Solution

$$y = \dfrac{x}{x-1}$$

Using quotient rule gives $\dfrac{dy}{dx}$ as follows:

$$\dfrac{dy}{dx} = \dfrac{(x-1)(1) - x(1)}{(x-1)^2}$$

$$= \dfrac{x-1-x}{(x-1)^2}$$

$$\dfrac{dy}{dx} = \dfrac{-1}{(x-1)^2}$$

Using chain rule gives $\dfrac{d^2y}{dx^2}$ as follows:

$$\dfrac{dy}{dx} = \dfrac{-1}{(x-1)^2}$$

$$= -1(x-1)^{-2}$$

$$\dfrac{d^2y}{dx^2} = -2 \text{ x } -1(x-1)^{-2-1} \text{ x } \dfrac{d(x)}{dx}$$

$$= 2(x-1)^{-3} \text{ x } 1$$

$$\dfrac{d^2y}{dx^2} = \dfrac{2}{(x-1)^3}$$

From the question, let us now simplify $(x-1)\dfrac{d^2y}{dx^2} + 2\dfrac{dy}{dx}$ as follows:

$$(x-1)\dfrac{d^2y}{dx^2} + 2\dfrac{dy}{dx}$$

$$= (x-1)\dfrac{2}{(x-1)^3} + (2)\dfrac{-1}{(x-1)^2}$$

$$= \dfrac{2}{(x-1)^2} - \dfrac{2}{(x-1)^2}$$

$$= 0$$

This shows that $(x-1)\dfrac{d^2y}{dx^2} + 2\dfrac{dy}{dx}$ is equal to zero.

15. Find, with respect to x, the derivative of $\left(x - \dfrac{5}{x}\right)^3$

Solution

$$y = \left(x - \frac{5}{x}\right)^3$$

This can also be written as:

$$y = (x - 5x^{-1})^3$$

$$\frac{dy}{dx} = 3(x - 5x^{-1})^{3-1} \times \frac{d(x - 5x^{-1})}{dx} \qquad \text{(By use of chain rule)}$$

$$= 3(x - 5x^{-1})^2 \times 1 - (-1 \times 5x^{-1-1})$$

$$= 3(x - 5x^{-1})^2 \times 1 - (-5x^{-2})$$

$$= 3(x - 5x^{-1})^2 \times (1 + 5x^{-2})$$

$$= 3(x - 5x^{-1})^2(1 + 5x^{-2})$$

$$\frac{dy}{dx} = 3\left(x - \frac{5}{x}\right)^2\left(1 + \frac{5}{x^2}\right)$$

16. Given that $\exp(2x^2 + 2y^2 - 16) = x + y$, find:

(a) $\dfrac{dy}{dx}$

(b) $\dfrac{dy}{dx}$ at $\left(\dfrac{1}{2}, \dfrac{1}{2}\right)$

Solution

$$\exp(2x^2 + 2y^2 - 16) = x + y$$

This can also be written as

$$e^{2x^2 + 2y^2 - 16} = x + y \qquad \text{(Note that } \exp x = e^x)$$

We now differentiate implicitly as follows:

$$\frac{d(2x^2 + 2y^2 - 16)}{dx} \times (e^{2x^2 + 2y^2 - 16}) = \frac{d(x)}{dx} + \frac{d(y)}{dx}$$

$$\left(4x + 4y\frac{dy}{dx}\right)(e^{2x^2 + 2y^2 - 16}) = 1 + \frac{dy}{dx}$$

Expanding the bracket gives:

$$4x(e^{2x^2 + 2y^2 - 16}) + 4y\frac{dy}{dx}(e^{2x^2 + 2y^2 - 16}) = 1 + \frac{dy}{dx}$$

Collecting terms in $\dfrac{dy}{dx}$ on the left hand side gives:

$$4y\frac{dy}{dx}(e^{2x^2 + 2y^2 - 16}) - \frac{dy}{dx} = 1 - 4x(e^{2x^2 + 2y^2 - 16})$$

Factorizing the left hand side gives:

$$\frac{dy}{dx}[4y(e^{2x^2 + 2y^2 - 16}) - 1] = 1 - 4x(e^{2x^2 + 2y^2 - 16})$$

Hence, $\dfrac{dy}{dx} = \dfrac{1 - 4x(e^{2x^2 + 2y^2 - 16})}{4y(e^{2x^2 + 2y^2 - 16}) - 1}$

Or, $\dfrac{dy}{dx} = \dfrac{-[4x(e^{2x^2 + 2y^2 - 16}) - 1]}{4y(e^{2x^2 + 2y^2 - 16}) - 1}$

(b) In order to find $\frac{dy}{dx}$ at $\left(\frac{1}{2} \ \frac{1}{2}\right)$, we simply substitute $x = \frac{1}{2}$ and $y = \frac{1}{2}$ into the expression for $\frac{dy}{dx}$ as follows:

$$\frac{dy}{dx} = \frac{-[4x(e^{2x^2+2y^2-16}) - 1]}{4y(e^{2x^2+2y^2-16}) - 1}$$

Since $x = \frac{1}{2}$ and $y = \frac{1}{2}$, this simplifies to give:

$$\frac{dy}{dx} \text{ at } \left(\frac{1}{2} \ \frac{1}{2}\right) = \frac{-[4\left(\frac{1}{2}\right)(e^{2x^2+2y^2-16}) - 1]}{4\left(\frac{1}{2}\right)(e^{2x^2+2y^2-16}) - 1}$$

$$= \frac{-[2(e^{2x^2+2y^2-16}) - 1]}{2(e^{2x^2+2y^2-16}) - 1}$$

The numerator cancels out the denominator to give:

$$\frac{dy}{dx} = \frac{-1}{1}$$

$$\frac{dy}{dx} = -1$$

17. If $y = x^3 - 2x^2 + 5$, show that: $x\frac{dy}{dx} - 3y - 2x^2 + 15 = 0$

Solution

$$y = x^3 - 2x^2 + 5$$

$$\frac{dy}{dx} = 3x^2 - 4x$$

Let us now simplify $x\frac{dy}{dx} - 3y - 2x^2 + 15$ as follows:

$x\frac{dy}{dx} - 3y - 2x^2 + 15$

$= x(3x^2 - 4x) - 3y - 2x^2 + 15$

$= x(3x^2 - 4x) - 3(x^3 - 2x^2 + 5) - 2x^2 + 15$ (Note that $x^3 - 2x^2 + 5$ has been substituted for y)

$= 3x^3 - 4x^2 - 3x^3 + 6x^2 - 15 - 2x^2 + 15$

$= 0$

Therefore, $x\frac{dy}{dx} - 3y - 2x^2 + 15 = 0$ as proven above.

18. Determine $\frac{d^2}{dx^2}\left(x\sin\frac{1}{x}\right)$

Solution

This means the second derivative of $x\sin\frac{1}{x}$

Let $y = x\sin\frac{1}{x}$

Or, $y = x\sin x^{-1}$

We now apply product rule to obtain $\dfrac{dy}{dx}$ as follows:

$$\frac{dy}{dx} = x\left[\frac{d(x^{-1})}{dx}\cos x^{-1}\right] + \sin x^{-1}\left[\frac{d(x)}{dx}\right]$$

$$= x(-x^{-2}\cos x^{-1}) + \sin x^{-1}(1)$$

$$= x(\frac{-1}{x^2}\cos x^{-1}) + \sin x^{-1}$$

$$= \frac{-x}{x^2}\cos x^{-1} + \sin x^{-1}$$

$$= \frac{-\cos x^{-1}}{x} + \sin x^{-1}$$

$$\frac{dy}{dx} = -\frac{1}{x}\cos\frac{1}{x} + \sin\frac{1}{x}$$

Or, $\dfrac{dy}{dx} = -x^{-1}\cos x^{-1} + \sin x^{-1}$

We now obtain $\dfrac{d^2y}{dx^2}$ as follows:

$$\frac{d^2y}{dx^2} = -x^{-1}\left[\frac{d(\cos x^{-1})}{dx}\right] + \cos x^{-1}\left[\frac{d(-x^{-1})}{dx}\right] + \frac{d(\sin x^{-1})}{dx}$$

$$= -x^{-1}\left[\frac{d(x^{-1})}{dx}(-\sin x^{-1})\right] + \cos x^{-1}(x^{-2}) + \left[\frac{d(x^{-1})}{dx}\cos x^{-1}\right]$$

$$= -x^{-1}[-x^{-2}(-\sin x^{-1})] + \cos x^{-1}(x^{-2}) - x^{-2}(\cos x^{-1})$$

$$= -x^{-1}(x^{-2}\sin x^{-1}) + x^{-2}\cos x^{-1} - x^{-2}\cos x^{-1}$$

$$= -x^{-1}(x^{-2}\sin x^{-1}) \quad \text{(Note that } x^{-2}\cos x^{-1} \text{ cancels out)}$$

$$= -x^{-3}\sin x^{-1}$$

$$\frac{d^2y}{dx^2} = -\frac{1}{x^3}\sin\frac{1}{x}$$

19. Given that $y = \dfrac{e^x + e^{-x}}{e^x - e^{-x}}$

(a) find $\dfrac{d^2y}{dx^2}$

(b) show that $\dfrac{d^2y}{dx^2} + 2y\dfrac{dy}{dx} = 0$

Solution

(a) $y = \dfrac{e^x + e^{-x}}{e^x - e^{-x}}$

By using quotient rule we obtain $\dfrac{dy}{dx}$ as follows:

$$\frac{dy}{dx} = \frac{e^x - e^{-x}\left[\frac{d(e^x + e^{-x})}{dx}\right] - \left(e^x + e^{-x}\left[\frac{d(e^x - e^{-x})}{dx}\right]\right)}{(e^x - e^{-x})^2}$$

$$= \frac{(e^x - e^{-x})(e^x - e^{-x}) - (e^x + e^{-x})(e^x + e^{-x})}{(e^x - e^{-x})^2}$$

$$= \frac{(e^x - e^{-x})^2 - (e^x + e^{-x})^2}{(e^x - e^{-x})^2}$$

Separating into fractions gives:

$$= \frac{(e^x - e^{-x})^2}{(e^x - e^{-x})^2} - \frac{(e^x + e^{-x})^2}{(e^x - e^{-x})^2}$$

$$\frac{dy}{dx} = 1 - \frac{(e^x + e^{-x})^2}{(e^x - e^{-x})^2}$$

We now apply quotient and chain rules to obtain $\frac{d^2y}{dx^2}$ as follows:

$$\frac{d^2y}{dx^2} = 0 - \frac{(e^x - e^{-x})^2[2(e^x + e^{-x})(e^x - e^{-x})] - (e^x + e^{-x})^2[2(e^x - e^{-x})(e^x + e^{-x})]}{[(e^x - e^{-x})^2]^2}$$

$$= -\left[\frac{2(e^x - e^{-x})^3(e^x + e^{-x}) - 2(e^x + e^{-x})^3(e^x - e^{-x})}{(e^x - e^{-x})^4}\right]$$

$$= \frac{-2(e^x - e^{-x})^3(e^x + e^{-x}) + 2(e^x + e^{-x})^3(e^x - e^{-x})}{(e^x - e^{-x})^4}$$

$$= \frac{2(e^x + e^{-x})^3(e^x - e^{-x}) - 2(e^x - e^{-x})^3(e^x + e^{-x})}{(e^x - e^{-x})^4} \qquad \text{(After rearranging the numerator)}$$

Separating into fractions gives:

$$= \frac{2(e^x + e^{-x})^3(e^x - e^{-x})}{(e^x - e^{-x})^4} - \frac{2(e^x - e^{-x})^3(e^x + e^{-x})}{(e^x - e^{-x})^4}$$

$$\frac{d^2y}{dx^2} = \frac{2(e^x + e^{-x})^3}{(e^x - e^{-x})^3} - \frac{2(e^x + e^{-x})}{e^x - e^{-x}}$$

(b) Let us simplify $\frac{d^2y}{dx^2} + 2y\frac{dy}{dx}$ as follows:

$$\frac{d^2y}{dx^2} + 2y\frac{dy}{dx}$$

$$= \frac{2(e^x + e^{-x})^3}{(e^x - e^{-x})^3} - \frac{2(e^x + e^{-x})}{e^x - e^{-x}} + 2\left(\frac{e^x + e^{-x}}{e^x - e^{-x}}\right)\left[1 - \frac{(e^x + e^{-x})^2}{(e^x - e^{-x})^2}\right]$$

Expanding bracket gives:

$$= \frac{2(e^x + e^{-x})^3}{(e^x - e^{-x})^3} - \frac{2(e^x + e^{-x})}{e^x - e^{-x}} + 2\left(\frac{e^x + e^{-x}}{e^x - e^{-x}}\right) - \frac{2(e^x + e^{-x})^3}{(e^x - e^{-x})^3}$$

$$= 0 \qquad \text{(Since equal terms with opposite signs cancel out each other)}$$

Therefore, $\frac{d^2y}{dx^2} + 2y\frac{dy}{dx}$ gives zero as proven above.

Or, $\frac{d^2y}{dx^2} + 2y\frac{dy}{dx} = 0$

20. Find the derivative of $\ln(\tan 2x)$

Solution

$y = \ln(\tan 2x)$

$$\frac{dy}{dx} = \frac{\frac{d(\tan 2x)}{dx}}{\tan 2x}$$

$$= \frac{2\sec^2 2x}{\tan 2x}$$

Or, $\dfrac{dy}{dx} = \dfrac{2\left(\frac{1}{\cos 2x}\right)\left(\frac{1}{\cos 2x}\right)}{\left(\frac{\sin 2x}{\cos 2x}\right)}$ (Note that $\sec 2x = \dfrac{1}{\cos 2x}$ and $\tan 2x = \dfrac{\sin 2x}{\cos 2x}$)

$$= \left(\frac{2}{\cos 2x}\right)\left(\frac{1}{\cos 2x}\right) \times \frac{\cos 2x}{\sin 2x}$$

$$\frac{dy}{dx} = \frac{2}{\sin 2x \cos 2x}$$ (Since $\cos 2x$ cancels $\cos 2x$)

21. Given that $y = x^3 + 3x^2$, determine $2\dfrac{dy}{dx} - x\dfrac{d^2y}{dx^2}$

Solution

$$y = x^3 + 3x^2$$

$$\frac{dy}{dx} = 3x^2 + 6x$$

$$\frac{d^2y}{dx^2} = 6x + 6$$

Hence, $2\dfrac{dy}{dx} - x\dfrac{d^2y}{dx^2}$ is simplified as follows:

$$2(3x^2 + 6x) - x(6x + 6)$$

$$= 6x^2 + 12x - 6x^2 - 6x$$

$$= 6x$$

Therefore, $2\dfrac{dy}{dx} - x\dfrac{d^2y}{dx^2} = 6x$

22. Find $\dfrac{dy}{dx}$ if $y = 3(3x + \sqrt{x})^2$

Solution

$$y = 3(3x + \sqrt{x})^2$$

Applying chain rule gives:

$$\frac{dy}{dx} = 3 \times 2 (3x + x^{\frac{1}{2}})^{2-1} \times \frac{d(3x + x^{\frac{1}{2}})}{dx}$$ (Note that $\sqrt{x} = x^{\frac{1}{2}}$)

$$= 6(3x + x^{\frac{1}{2}}) \times (3 + \frac{1}{2}x^{-\frac{1}{2}})$$

$$= 6(3x + x^{\frac{1}{2}})\left(3 + \frac{1}{2x^{\frac{1}{2}}}\right)$$

$$= 6(3x + \sqrt{x})\left(3 + \frac{1}{2\sqrt{x}}\right)$$

Expanding the bracket gives:

$$= 6\left(9x + \frac{3x}{2\sqrt{x}} + 3\sqrt{x} + \frac{1}{2}\right)$$

$$= 6\left(9x + \frac{3}{2}\sqrt{x} + 3\sqrt{x} + \frac{1}{2}\right) \qquad \text{(Note that } \frac{3x}{2\sqrt{x}} = \frac{3}{2}x^{1-\frac{1}{2}} = \frac{3}{2}x^{\frac{1}{2}} = 3\sqrt{x}\text{)}$$

$$= 6\left(9x + \frac{9}{2}\sqrt{x} + \frac{1}{2}\right)$$

$$\frac{dy}{dx} = 54x + 27\sqrt{x} + 3$$

23. If $y = \dfrac{5x^2 + 7}{x^4}$,

(a) find $\dfrac{d^2y}{dx^2}$

(b) show that $x^2\dfrac{d^2y}{dx^2} + 7x\dfrac{dy}{dx} + 8y = 0$

Solution

(a) $y = \dfrac{5x^2 + 7}{x^4}$

$$= \frac{5x^2}{x^4} + \frac{7}{x^4}$$

$$= \frac{5}{x^2} + \frac{7}{x^4}$$

$$y = 5x^{-2} + 7x^{-4}$$

$$\frac{dy}{dx} = -10x^{-3} - 28x^{-5}$$

Similarly,

$$\frac{d^2y}{dx^2} = 30x^{-4} + 140x^{-6}$$

(b) Let us now simplify $x^2\dfrac{d^2y}{dx^2} + 7x\dfrac{dy}{dx} + 8y$ by substituting appropriately as follows:

$$x^2(30x^{-4} + 140x^{-6}) + 7x(-10x^{-3} - 28x^{-5}) + 8(5x^{-2} + 7x^{-4})$$

$$= 30x^{-2} + 140x^{-4} - 70x^{-2} - 196x^{-4} + 40x^{-2} + 56x^{-4}$$

$$= 30x^{-2} + 40x^{-2} - 70x^{-2} + 140x^{-4} + 56x^{-4} - 196x^{-4}$$

$$= 0$$

Therefore $x^2\dfrac{d^2y}{dx^2} + 7x\dfrac{dy}{dx} + 8y$ gives zero as obtained above.

Or, $x^2\dfrac{d^2y}{dx^2} + 7x\dfrac{dy}{dx} + 8y = 0$

24. If $y = (2x + 5)^4 + \dfrac{x - 1}{2x - 1}$ find $\dfrac{dy}{dx}$.

Solution

$$(2x + 5)^4 + \frac{x - 1}{2x - 1}$$

We now use chain rule and quotient rule as follows:

$$\frac{dy}{dx} = 4(2x+5)^{4-1} \times \frac{d(2x+5)}{dx} + \frac{(2x-1)\frac{d(x-1)}{dx} - \left[(x-1)\frac{d(2x-1)}{dx}\right]}{(2x-1)^2}$$

$$= 4(2x+5)^3(2) + \frac{(2x-1)(1) - [(x-1)(2)]}{(2x-1)^2}$$

$$= 8(2x+5)^3 + \frac{(2x-1) - 2(x-1)}{(2x-1)^2}$$

$$= 8(2x+5)^3 + \frac{2x-1-2x+2}{(2x-1)^2}$$

$$\frac{dy}{dx} = 8(2x+5)^3 + \frac{1}{(2x-1)^2}$$

Exercise 17

1. If $y = x^2(3x^4 - 5)^3$ find $\frac{dy}{dx}$.

2. Given that $f(x) = 2x^5 - x^4 + 2x^3 + x^2 - 3x + 4$ find $f'(-2)$

3. A function $f(x)$ is given by $f(x) = \frac{\sqrt{(3x^2 - 1)^3}}{x^2}$. Find:

(a) the derivative of $f(x)$

(b) the gradient of $f(x)$ at the point $(1, -2)$

4. If $3xy^3 - y^2 - 4x^3 = 5y$, find:

(a) the derivative of the expression

(b) the gradient at $(-1, 1)$

5. Find the derivative of $(2x^2 + y^2)^2 = 0$

6. If $y = (x^2 + 5)^3(x^3 - 1)$, find $\frac{dy}{dx}$.

7. Find the derivative of $e^{\sin x + \tan x}$

8. Find the derivative of $\frac{x^2}{\sin^2 x}$

9. Differentiate with respect to x: $\ln(\sin^2 x + \cos 3x)$

10. If $y = a^{\cos 5x}$ find $\frac{dy}{dx}$.

11. Given that $y = x^2 e^{-3x}$ evaluate $\frac{d^2y}{dx^2} + \frac{dy}{dx} - y$

12. Differentiate with respect to x: $\sin\left(\frac{1 - 2x^3}{x^2}\right)$

13. If $y = \frac{2x-1}{x^2}$, find $x^4\frac{d^2y}{dx^2} - 3(2x-1)\frac{dy}{dx}$

147

14. Given that $y = 5e^{-2x} + 3e^x$, evaluate $\dfrac{d^2y}{dx^2} + \dfrac{dy}{dx} - 2y$

15. Find the derivative of $\left(\dfrac{1}{x^2} - \dfrac{2}{x}\right)^5$

16. Given that $e^{x^3 - y^3} = xy$, find:

(a) $\dfrac{dy}{dx}$

(b) $\dfrac{dy}{dx}$ at $(1, -1)$

17. If $y = \sin(\sin x)$, evaluate $\dfrac{d^2y}{dx^2} + \tan x \dfrac{dy}{dx} + y\cos^2 x$

18. Determine $\dfrac{d^2}{dx^2}\left(x^2 \cos\dfrac{1}{x^2}\right)$

19. Given that $y = \dfrac{1 + e^{-x}}{1 - e^{-x}}$

(a) find $\dfrac{d^2y}{dx^2}$

(b) evaluate $\dfrac{d^2y}{dx^2} - \dfrac{dy}{dx} + y$

20. Find the derivative of $\ln(\sec^2 x)$

21. Given that $y = x + \tan x$, determine $\cos^2 x \dfrac{d^2y}{dx^2} - 2y + 2x$

22. Find $\dfrac{dy}{dx}$ if $y = \left(3x + \dfrac{\sqrt{5x}}{2}\right)^3$

23. If $y = \dfrac{x^3 + 2x^2 - 5x - 3}{x^2}$, find $\dfrac{d^2y}{dx^2}$

24. If $y = (x^3 + 1)^3 + \dfrac{5x - 2}{x^2 + 1}$ find $\dfrac{dy}{dx}$.

25. Given that $y = 2\cos 5x + 7\sin 5x$, determine $\dfrac{d^2y}{dx^2} + 25y$

CHAPTER 18
COLLECTION AND TABULATION OF DATA

When a large volume of data is obtained, it is necessary to present such data in frequency table. Sometimes the tally system which involves the use of vertical and horizontal strokes is applied.

Examples

1. A die is rolled 50 times and the following data is obtained. Represent the data in a frequency table.

4	6	4	3	5	3	1	4	6	5	6	4	2
6	4	5	6	2	1	6	4	3	4	6	1	5
1	3	6	2	2	4	3	4	5	3	4	1	2
3	1	2	1	5	3	4	3	4	2	5		

Solutions

The data which ranges from 1 to 6 is summarized as shown on the table below.

Number on die	Frequency
1	7
2	7
3	9
4	12
5	7
6	8

The data can also be represented on a horizontal table as shown below.

Number in die	1	2	3	4	5	6
Frequency	7	7	9	12	7	8

2. The scores of 40 students in a physics test are presented below. Prepare a frequency distribution table for the data.

64	66	68	63	70	63	67	64	70	69
66	64	65	70	62	70	66	69	67	64
61	63	67	62	68	64	63	69	70	63
63	61	68	67	68	63	61	67	69	68

150

<u>Solution</u>

The data which ranges from 61 to 70 is summarized as shown on the table below.

Score	61	62	63	64	65	66	67	68	69	70
Frequency	3	2	7	5	1	3	5	5	4	5

Exercise 18

1. The marks obtained in an examination by 40 students in a class are as shown below. Represent the data in a frequency table.

71	74	74	70	70	72	74	74	65	69
66	68	65	73	66	72	69	69	67	65
71	73	67	68	68	69	70	69	71	65
67	67	68	72	74	73	67	67	69	68

2. The score of 20 students in a test are as shown below. Represent the score in a frequency table.

6	6	8	9	5	6	7	5	6	9
8	6	7	5	9	5	9	6	6	5

3. The number of seeds in a sample of 40 cocoa pods are as shown below. Represent the information using a frequency table.

28	22	28	28	27	29	20	20	20	24
21	25	25	20	22	20	26	29	30	24
21	23	27	22	28	30	23	29	20	23
23	21	28	27	28	23	21	27	30	28

4. The ages of 30 students in a senior high school is represented below. Show the data using a frequency table.

14	15	12	13	10	13	11	14	10	12
15	11	15	10	12	10	11	13	14	14
15	13	12	11	14	14	13	12	10	15

CHAPTER 19
MEAN, MEDIAN AND MODE OF UNGROUPED DATA

The mean, median and mode are averages of sets of statistical data. They are called "measures of central tendency".

Mean

The mean of a set of data is obtained by adding all the data and then dividing the result by the total number in the data set.

For an ungrouped data given in a frequency table, the mean can be calculated by using the formula:

$$\bar{x} = \frac{\sum fx}{\sum f}$$

Where $\bar{x}$ is the mean, $\sum$ is a symbol representing summation, x is each number in the data, and f is the frequency.

Median

The median of a data is the middle number when the data is arranged in an increasing or decreasing order of size. For an odd number of data, the position of the middle number is obtained by the expression:

$$\text{Median} = \text{number in the } \left(\frac{N+1}{2}\right)\text{th position}$$

where N is the total number of data.

If there is an even number of data, the median is the average of the two middle numbers. In such a case the positions of the two middle numbers is obtained by the expression:

$$\text{Median} = \frac{\text{Number in the } \left(\frac{N}{2}\right)\text{th position} + \text{Number in the } \left(\frac{N+2}{2}\right)\text{th position}}{2}$$

However, for large data in a frequency table which has an even number of data, the median is given by:

$$\text{Median} = \frac{\text{Number in the } \left(\frac{\sum f}{2}\right)\text{th position} + \text{Number in the } \left(\frac{\sum f + 2}{2}\right)\text{th position}}{2}$$

Where $\sum f$ is the total frequency of the data.

Note that for an odd number of data, only one number will be at the middle. However, for an even number of data, two numbers will be at the middle. The average of the two numbers gives the median of the data.

Mode

The mode is the most occurring number in a set of data. It is the number with the highest frequency. If a set of data has two modes, we say it is bimodal.

Examples

1. Find the mean, median and mode of the data below:

  2, 5, 0, 3, 1, 6, 9, 7, 3

Solution

There are 9 numbers in the data. So, the mean is obtained as follows:

$$\text{Mean} = \frac{2+5+0+3+1+6+9+7+3}{9}$$

$$= \frac{36}{9}$$

$$= 4$$

∴   The mean is 4

In order to calculate the median, first arrange the numbers in ascending order as follows:

  0, 1, 2, 3, 3, 5, 6, 7, 9

By inspection, the number that is at the middle of the data is 3

∴   The median is 3

Or, since there are 9 numbers in the data and 9 is an odd number, then the position of the middle number is obtained as follows:

$$\text{Median} = \text{number in the } \left(\frac{N+1}{2}\right)\text{th position}$$

$$= \text{number in the } \left(\frac{9+1}{2}\right)\text{th position}$$

$$= \text{number in the } \left(\frac{10}{2}\right)\text{th position}$$

$$= \text{number in the } 5^{th} \text{ position.}$$

$$= 3 \text{ (Since 3 is in the } 5^{th} \text{ position in the data arranged above)}$$

∴ The median is 3

The most occurring number in the data is 3 since it appears twice while every other number appears once.

∴ The mode is 3.

2. Find: a. the mean;

  b. the median;

  c. the mode of the data below.

  6, 7, 10, 5, 11, 5, 9, 7, 10, 13, 5, 8, 7, 5, 12

Solutions

a. There are 15 numbers in the data. So, the mean is obtained as follows:

$$\text{Mean} = \frac{6 + 7 + 10 + 5 + 11 + 5 + 9 + 7 + 10 + 13 + 5 + 8 + 7 + 5 + 12}{15}$$

$$= \frac{120}{15}$$

$$= 8$$

∴   The mean is 8

b. In order to calculate the median, first arrange the numbers in ascending order as follows:

5, 5, 5, 5, 6, 7, 7, 7, 8, 9, 10, 10, 11, 12, 13

By inspection, the number that is at the middle of the data is 7

∴   The median is 7

Or, since there are 15 numbers in the data and 15 is an odd number, then the position of the middle number is obtained as follows:

$$\text{Median} = \text{number in the } \left(\frac{N+1}{2}\right)\text{th position}$$

$$= \text{number in the } \left(\frac{15+1}{2}\right)\text{th position}$$

$$= \text{number in the } \left(\frac{16}{2}\right)\text{th position}$$

$$= \text{number in the 8}^{\text{th}} \text{ position.}$$

$$= 7 \quad (\text{Since 7 is in the 8}^{\text{th}} \text{ position of the data})$$

∴   The median is 7

c. The most occurring number in the data is 5. It occurs four times.

∴   The mode is 5.

3. Find the mean, median and mode of the data below:

11, 14, 10, 16, 18, 12, 11, 15, 10, 11, 15, 13

Solutions

There are 12 numbers in the data. So, the mean is obtained as follows:

$$\text{Mean} = \frac{11 + 14 + 10 + 16 + 18 + 12 + 11 + 15 + 10 + 11 + 15 + 13}{12}$$

$$= \frac{156}{12}$$

$$= 13$$

∴ The mean is 13

In order to calculate the median, first arrange the numbers in ascending order as follows:

10, 10, 11, 11, 11, 12, 13, 14, 15, 15, 16, 18

By inspection, the two numbers that are at the middle of the data are 12 and 13. So, we take their average.

$$median = \frac{12+13}{2}$$

$$= \frac{25}{2}$$

$$= 12.5$$

∴ The median is 12.5

Or, since there are 12 numbers in the data and 12 is an even number, then the positions of the two middle numbers and their average are obtained as follows:

$$Median = \frac{\text{Number in the } \left(\frac{N}{2}\right)\text{th position} + \text{Number in the } \left(\frac{N+2}{2}\right)\text{th position}}{2}$$

$$= \frac{\text{Number in the } \left(\frac{12}{2}\right)\text{th position} + \text{Number in the } \left(\frac{12+2}{2}\right)\text{th position}}{2}$$

$$= \frac{\text{Number in the 6th position} + \text{Number in the } \left(\frac{14}{2}\right)\text{th position}}{2}$$

$$= \frac{12+13}{2} \quad \text{(Note that 12 is in the 6}^{\text{th}}\text{ position while 13 is in the 7}^{\text{th}}\text{ position)}$$

$$= \frac{25}{2}$$

$$= 12.5$$

∴ The median is 12.5

The most occurring number in the data is 11 since it appears three times.

∴ The mode is 11.

4. Find the mean, median and mode of the data below:

    50, 56, 58, 52, 55, 59, 51, 55

Solutions

There are 8 numbers in the data. So, the mean is obtained as follows:

$$Mean = \frac{50 + 56 + 58 + 52 + 55 + 59 + 51 + 55}{8}$$

$$= \frac{436}{8}$$

$$= 54.5$$

∴ The mean is 54.5

In order to calculate the median, first arrange the numbers in ascending order as follows:

50, 51, 52, 55, 55, 56, 58, 59.

By inspection, the two numbers that are at the middle of the data are 55 and 55. So, we take their average.

$$\text{median} = \frac{55+55}{2}$$

$$= \frac{110}{2}$$

$$= 55$$

∴ The median is 55

Or, since there are 8 numbers in the data and 8 is an even number, then the positions of the two middle numbers and their average are obtained as follows:

$$\text{Median} = \frac{\text{Number in the } \left(\frac{N}{2}\right)\text{th position} + \text{Number in the } \left(\frac{N+2}{2}\right)\text{th position}}{2}$$

$$= \frac{\text{Number in the } \left(\frac{8}{2}\right)\text{th position} + \text{Number in the } \left(\frac{8+2}{2}\right)\text{th position}}{2}$$

$$= \frac{\text{Number in the 4th position} + \text{Number in the } \left(\frac{10}{2}\right)\text{th position}}{2}$$

$$= \frac{\text{Number in the 4th position} + \text{Number in the 5th position}}{2}$$

$$= \frac{55 + 55}{2} \quad \text{(Note that 55 is in the 4}^{th}\text{ and the 5}^{th}\text{ position)}$$

$$= \frac{110}{2}$$

$$= 55$$

∴ The median is 55

The most occurring number in the data is 55.

∴ The mode is 55.

5. The table below shows the marks of 50 students in a test.

Mark	1	2	3	4	5	6	7
No of Student	8	16	10	5	3	6	2

Calculate: a. the mean b. the median c. the mode of the marks

Solutions

a. Note that the number of students is also the frequency. Presenting the table as shown

156

below allows for easy calculation of the mean.

Mark (x)	No of student (f)	Fx
1	8	8
2	16	32
3	10	30
4	5	20
5	3	15
6	6	36
7	2	14
Total:	$\sum f = 50$	$\sum fx = 155$

Note that the column fx is obtained by multiplying the values of numbers in the column f by numbers in the column x.

Mean, $\bar{x} = \dfrac{\sum fx}{\sum f}$

$= \dfrac{155}{50}$

$= 3.1$

b. Since there are 50 students, i.e. the total frequency is 50, and 50 is an even number, then the positions of the two middle marks and their average are obtained as follows:

$$\text{Median} = \frac{\text{Number in the } \left(\frac{\sum f}{2}\right)\text{th position} + \text{Number in the } \left(\frac{\sum f + 2}{2}\right)\text{th position}}{2}$$

$$= \frac{\text{Number in the } \left(\frac{50}{2}\right)\text{th position} + \text{Number in the } \left(\frac{50+2}{2}\right)\text{th position}}{2}$$

$$= \frac{\text{Number in the 25th position} + \text{Number in the } \left(\frac{52}{2}\right)\text{th position}}{2}$$

$$= \frac{\text{Number in the 25th position} + \text{Number in the 26th position}}{2}$$

$$= \frac{3 + 3}{2} \quad \text{(Note that 3 is the mark in the 25}^{\text{th}} \text{ position and in the 26}^{\text{th}} \text{ position)}$$

$$= \frac{6}{2}$$

$$= 3$$

$\therefore$ The median is 3

Use the frequency (number of students) to locate the marks in the 25th and 26th position as follows:

The first frequency of 8 shows that mark 1 occupies the position of 1st to 8th. Adding the second frequency of 16 to the first frequency gives, 8 + 16 = 24. This shows that after the 8th position occupied by the mark 1, the positions 9th to 24th is occupied by the mark 2. Adding the third frequency of 10 to the previous sum of frequencies gives, 10 + 24 = 34. This shows that after the 24th position occupied by the mark 2, the positions 25th to 34th is occupied by the mark 3. Hence, 3 is the mark in the 25th and 26th position which are at the middle of the data.

c. The mode is the mark that has the highest frequency. From the table, the mark 2 has the highest frequency of 16. So, the mode is 2.

∴ The mode is 2.

Note that the mode is not the frequency itself, but that particular mark that has the highest frequency. Avoid the mistake of taking 16 (frequency) as the mode.

6. The table below shows the ages of 30 students in a school.

Age	10	11	12	13	14	15
No of Student	1	4	3	7	9	6

Calculate: a. the mean b. the median c. the mode of the ages

Solutions

a. Using the number of students as the frequency, the table can be presented for easy calculation of the mean as follows:

Age (x)	No of student (f)	Fx
10	1	10
11	4	44
12	3	36
13	7	91
14	9	126
15	6	90
Total:	$\sum f = 30$	$\sum fx = 397$

Note that the column fx is obtained by multiplying the values of numbers in the column f by numbers in the column x.

$$\text{Mean, } \bar{x} = \frac{\Sigma fx}{\Sigma f}$$

$$= \frac{397}{30}$$

$$= 13.3$$

b. Since there are 30 students, i.e. the total frequency is 30, and 30 is an even number, then the positions of the two middle ages and their average are obtained as follows:

$$\text{Median} = \frac{\text{Age in the } \left(\frac{\Sigma f}{2}\right)\text{th position} + \text{Age in the } \left(\frac{\Sigma f + 2}{2}\right)\text{th position}}{2}$$

$$= \frac{\text{Age in the } \left(\frac{30}{2}\right)\text{th position} + \text{Age in the } \left(\frac{30+2}{2}\right)\text{th position}}{2}$$

$$= \frac{\text{Age in the 15th position} + \text{Age in the } \left(\frac{32}{2}\right)\text{th position}}{2}$$

$$= \frac{\text{Age in the 15th position} + \text{Age in the 16th position}}{2}$$

$$= \frac{13 + 14}{2}$$

$$= \frac{27}{2}$$

$$= 13.5$$

∴   The median is 13.5

Note that 13 is the age in the 15th position while 14 is the age in the 16th position.

The frequency (number of students) was used to locate the ages in the 15th and 16th position as follows:

The first frequency of 1 shows that age 10 occupies the 1st position. Adding the second frequency of 4 to the first frequency gives, 1 + 4 = 5. This shows that after the 1th position occupied by the age 10, the positions 2nd to 5th is occupied by the age 11. Adding the third frequency of 3 to the previous sum of frequencies gives, 3 + 5 = 8. This shows that after the 5th position occupied by the age 11, the positions 6th to 8th is occupied by the age 12. Adding the fourth frequency of 7 to the previous sum of frequencies gives, 7 + 8 = 15. This shows that after the 8th position occupied by the age 12, the positions 9th to 15th is occupied by the age 13. Adding the fifth frequency of 9 to the previous sum of frequencies gives, 9 + 15 = 24. This shows that after the 15th position occupied by the age 13, the positions 16th to 24th is occupied by the age 14. Hence, 13 is the age in the 15th position while 14 is the age in the 16th position.

c.   The mode is the age that has the highest frequency. From the table, the age 14 has the highest frequency of 9.

∴ The mode is 14.

Note that 9 is the frequency. It should not be taken as the mode.

Range

Range is the difference between the highest and lowest values in a given set of data. It is a measure of dispersion or variation.

Examples

1. Find the range of the following set of numbers: 4, 8, 2, 5, 8, 3, 6, 4, 9, 2, 5

<u>Solution</u>

The highest number in the data set is 9, while the lowest number is 2.
∴   Range = Highest number – Lowest number
         = 9 – 2 = 7
     Range = 7

2. The monthly salaries of five workers in a company are: $845, $1205, $694, $626 and $864. What is the range of the salaries?

<u>Solution</u>

Range = Highest salary – Lowest salary
         = 1205 – 626
   Range = $579

Exercise 19

1. Find the mean, median and mode of the data below:
        1, 6, 10, 4, 1, 2, 5, 2, 3, 2, 8
2. Find: a. the mean;
          b. the median;
          c. the mode of the data below.
     20, 24, 21, 25, 22, 25, 28, 26, 20, 23, 25, 27 and 26
3. Find the mean, median and mode of the data below:
      101, 105, 120, 116, 109, 112, 118, 115, 105 and 111
4. Find the mean, median and mode of the data below:

0, 6, 8, 2, 5, 9, 1, 5, 4, 7, 5, 2, 3, 3

5. The table below shows the marks of 50 students in a test.

Mark	3	4	5	6	7	8	9
No of Student	8	16	10	5	3	6	2

Calculate: a. the mean b. the median c. the mode of the marks

6. The table below shows the ages of 30 students in a school.

Age	10	11	12	13	14	15
No of Student	1	4	3	7	9	6

Calculate: a. the mean b. the median c. the mode of the ages

7. Find the range of the following set of data

a. 12, 17, 21, 15, 19, 13, 11, 16, 22, 12, 13

b. 231kg, 258kg, 213kg, 243kg, 216kg, 271kg, 262kg, 219kg, 238kg, 231kg.

CHAPTER 20
COLLECTION AND TABULATION OF GROUPED DATA

Statistical data containing numerous values is easier to work with when the values are grouped into class intervals.

Examples

1. The data below gives the marks of 30 students in an exam.

43	45	50	47	51	58	52	47	42	54
61	50	45	55	57	41	46	49	51	50
59	44	53	57	49	40	48	52	51	58

Taking class intervals 40 – 44, 45 – 49, ..., construct a frequency distribution for the data.

Solution

The data is summarized as shown on the table below. Note that the highest value in the given data falls within the range 60 – 64.

Class interval	40 – 44	45 - 49	50 - 54	55 - 59	60 - 64
Frequency	5	8	10	6	1

2. The data below gives the ages of lecturers in a university.

34	62	54	41	51	63	31	44	48	50
33	45	59	55	47	31	39	55	60	40
63	45	53	55	36	58	61	34	34	43
47	35	43	51	35	48	42	51	36	31

Taking class intervals 31 – 35, 36 – 40, ..., construct a frequency table for the data.

Solution

The data is summarized as shown on the table below. Note that the highest value in the given data falls within the range 61 – 65.

Age	31 – 35	36 - 40	41 - 45	46 - 50	51 - 55	56 - 60	61 - 65
Frequency	9	4	7	5	8	3	4

Terms Used in Grouped Data

The table below will be used to explain the terms used in grouped data.

Class interval	Frequency
8 – 14	3
15 – 21	5
22 – 28	8
29 – 35	18

1. Class limit: The end numbers in each class interval are called the class limits. 8 is the lower class limit, while 14 is the upper class limit of the first class interval.

2. Class boundaries: The class boundary for the second class interval is 14.5 – 21.5. The lower class boundary is 14.5 which is obtained by subtracting 0.5 from 15 (the lower class limit). The upper class boundary is 21.5 which is obtained by adding 0.5 to 21 (the upper class limit). Other class boundaries are obtained in a similar way.

3. Class width: For each class interval the difference between the upper class boundary and the lower class boundary gives the class width. From the table above the class width for the third class interval is 28.5 – 21.5 = 7

4. Class mid-value: This is half of the sum of the lower and upper class limit of a given class interval. The class-value of the first class interval is given by: $\frac{8+14}{2} = \frac{22}{2} = 11$.

Examples

1. Copy and complete the table below.

Class interval	Frequency	Class boundary	Class width	Class mid-value
55 – 59	3			
60 – 64	2			
65 – 69	5			
70 – 74	4			
75 – 79	1			

<u>Solution</u>

The completed table is as shown below

Class interval	Frequency	Class boundary	Class width	Class mid-value
55 – 59	3	54.5 – 59.5	5	57
60 – 64	2	59.5 – 64.5	5	62
65 – 69	5	64.5 – 69.5	5	67
70 – 74	4	69.5 – 74.5	5	72
75 – 79	1	74.5 – 79.5	5	77

Note that the class boundaries are obtained by subtracting and adding 0.5 to the lower and upper class limits respectively. This 0.5 is obtained by finding the difference between the lower class limit of one class and the upper class limit of the previous class and dividing the result by 2. This gives, for example, $\frac{60-59}{2} = \frac{1}{2} = 0.5$.

The class width is the difference between the upper and lower class boundaries. The class mid-values are obtained by taking the mean of the upper and lower class limits.

2. Copy and complete the table below.

Class interval	Frequency	Class boundary	Class width	Class mid-value
0 – 19	2			
20 – 39	8			
40 – 59	3			
60 – 79	1			
80 – 99	4			

Solution

The completed table is as shown below

Class interval	Frequency	Class boundary	Class width	Class mid-value
0 – 19	2	-0.5 – 19.5	20	9.5
20 – 39	8	19.5 – 39.5	20	29.5
40 – 59	3	39.5 – 59.5	20	49.5
60 – 79	1	59.5 – 79.5	20	69.5
80 – 99	4	79.5 – 99.5	20	89.5

Exercise 20

1. The data below gives the scores of 50 students in an exam.

43	45	50	47	51	58	52	47	42	54
61	50	45	55	57	41	46	49	51	50
59	44	53	57	49	40	48	52	51	58
48	54	43	54	61	60	49	57	45	42
56	45	57	61	54	62	44	47	46	62

Taking class intervals 40 – 44, 45 – 49, …, construct a frequency distribution for the data.

2. The data below shows the weights in kg of students in a school.

24	32	44	51	31	23	51	34	48	40
53	45	29	35	27	51	29	35	50	30
43	55	53	35	26	28	41	44	54	43
27	45	33	41	55	28	32	51	26	39

Taking class intervals 21 – 25, 26 – 30, …, construct a frequency table for the data.

3. Copy and complete the table below.

Class interval	Frequency	Class boundary	Class width	Class mid-value
5 – 9	2			
10 – 14	5			
15 – 19	5			
20 – 24	7			
25 – 29	1			

4. Copy and complete the table below.

Class interval	Frequency	Class boundary	Class width	Class mid-value
1 – 20	1			
21 – 40	4			
41 – 60	7			
61 – 80	3			
81 – 100	5			

5. Copy and complete the table below.

Class interval	Frequency	Class boundary	Class width	Class mid-value
0 – 90	2			
100 – 190	4			
200 – 290	1			
300 – 390	7			
400 – 490	1			

CHAPTER 21
MEAN, MEDIAN AND MODE OF GROUPED DATA

Mean

The mean of a grouped data can be calculated by substituting the class mid value as the values of x in the formula given by:

$$\text{Mean } \bar{x} = \frac{\Sigma fx}{\Sigma f}$$

Median

The median of a grouped data can be estimated by:

$$\text{Median} = L + C\left(\frac{\frac{\Sigma f}{2} - CF_{bm}}{F_m}\right)$$

Where, $\frac{\Sigma f}{2}$ determines the median class

L = Lower class boundary of the median class

CF_{bm} = Cumulative frequency before the median class

F_{m} = Frequency of the median class

C = Class width

Mode

The mode of a grouped data can be calculated as follows:

$$\text{Mode} = L + C\left(\frac{\Delta_1}{\Delta_1 + \Delta_2}\right)$$

Where, L = Lower class boundary of modal class

C = Class width

Δ_1 = Difference between the frequency of the modal class and the frequency before it

Δ_2 = Difference between the frequency of the modal class and the frequency after it

Examples

1. The following table shows the frequency distribution of ages, in years of 50 people at a bus stop.

Ages	10 - 19	20 - 29	30 - 39	40 - 49	50 - 59	60 - 69
Number of people	6	12	16	9	5	2

Calculate: a. the mean

b. the median

c. the mode of the distribution

Solution

Ages	Number of people (f)	Cumulative frequency	Class mid-value(x)	fx	Class boundary	Class width
10 – 19	6	6	14.5	87	9.5-19.5	10
20 – 29	12	6+12=18	24.5	294	19.5-29.5	10
30 – 39	16	18+16=34	34.5	552	29.5-39.5	10
40 – 49	9	34+9=43	44.5	400.5	39.5-49.5	10
50 – 59	5	43+5=48	54.5	272.5	49.5-59.5	10
60 – 69	2	48+2=50	64.5	129	59.5-69.5	10
	$\sum f = 50$			$\sum f = 1735$		

a. Mean $\bar{x} = \dfrac{\sum fx}{\sum f}$

$\quad = \dfrac{1735}{50} = 34.7$

b. Median $= L + C\left(\dfrac{\frac{\sum f}{2} - CF_{bm}}{F_m}\right)$

$\dfrac{\sum f}{2} = \dfrac{50}{2} = 25$

This shows that the median class falls in the 25th position

∴   Median class = 30 – 39

Lower class boundary of median class, L = 29.5

Class width, C = 10

Cumulative frequency before the median class, CF_{bm} = 18

Frequency of the median class, F_m = 16

∴   Median $= L + C\left(\dfrac{\frac{\sum f}{2} - CF_{bm}}{F_m}\right)$

$\quad = 29.5 + 10\left(\dfrac{25 - 18}{16}\right)$

$\quad = 29.5 + 10\left(\dfrac{7}{16}\right)$

$\quad = 29.5 + 4.375 = 33.875$

∴   Median = 33.9 (To 1 d.p)

c. Mode $= L + C\left(\dfrac{\Delta_1}{\Delta_1 + \Delta_2}\right)$

Modal class = 30 – 39

Lower class boundary of modal class, L = 29.5

Class width = 10

Δ_1 = Modal class frequency – frequency before it

    = 16 - 12 = 4

Δ_2 = Modal class frequency – frequency after it

    = 16 - 9 = 7

$\therefore$ Mode = L + C$\left(\dfrac{\Delta_1}{\Delta_1 + \Delta_2}\right)$

    = 29.5 + 10$\left(\dfrac{4}{4+7}\right)$

    = 29.5 + 10$\left(\dfrac{4}{11}\right)$

    = 29.5 + 3.64 = 33.14

$\therefore$ Mode = 33.1 (To 1 d.p)

2. The data below is the weight of students in a high school.

Weight	31 - 35	36 - 40	41 - 45	46 - 50	51 - 55	56 - 60	61 - 65
Number of student	2	9	7	5	8	3	6

Determine: a. the mean

b. the median

c. the mode of the weights

Solution

Weight	Number of student (f)	Cumulative frequency	Class mid-value(x)	fx	Class boundary	Class width
31 – 35	2	2	33	66	30.5-35.5	5
36 – 40	9	2+9=11	38	349	35.5-40.5	5
41 – 45	7	11+7=18	43	301	40.5-45.5	5
46 – 50	5	18+5=23	48	240	45.5-50.5	5
51 – 55	8	23+8=31	53	424	50.5-55.5	5
56 – 60	3	31+3=34	58	174	55.5-60.5	5
61 – 65	6	34+6=40	63	378	60.5-65.5	5
	Σf = 40			Σfx= 1932		

a. Mean $\bar{x} = \dfrac{\Sigma fx}{\Sigma f}$

$$= \frac{1932}{40}$$

$$= 48.3$$

b.　Median $= L + C\left(\dfrac{\frac{\Sigma f}{2} - CF_{bm}}{F_m}\right)$

$$\frac{\Sigma f}{2} = \frac{40}{2} = 20$$

This shows that the median class falls in the 20th position

∴　Median class $= 46 - 50$

Lower class boundary of median class, $L = 45.5$

Class width, $C = 5$

Cumulative frequency before the median class, $CF_{bm} = 18$

Frequency of the median class, $F_m = 5$

∴　Median $= L + C\left(\dfrac{\frac{\Sigma f}{2} - CF_{bm}}{F_m}\right)$

$$= 45.5 + 5\left(\frac{20 - 18}{5}\right)$$

$$= 45.5 + 5\left(\frac{2}{5}\right)$$

$$= 45.5 + 2 = 47.5$$

∴　Median $= 47.5$

c.　Mode $= L + C\left(\dfrac{\Delta_1}{\Delta_1 + \Delta_2}\right)$

Modal class $= 36 - 40$

Lower class boundary of modal class, $L = 35.5$

Class width $= 5$

Δ_1 = Modal class frequency – frequency before it

　　$= 9 - 2 = 7$

Δ_2 = Modal class frequency – frequency after it

　　$= 9 - 7 = 2$

∴　Mode $= L + C\left(\dfrac{\Delta_1}{\Delta_1 + \Delta_2}\right)$

$$= 35.5 + 5\left(\frac{7}{7 + 2}\right)$$

$$= 35.5 + 5\left(\frac{7}{9}\right)$$

$$= 35.5 + 3.89 = 39.39$$

∴　Mode $= 39.4$　(To 1 d.p)

Exercise 21

1. The following table shows the weights of 30 people at a company.

Weight	60 - 64	65 - 69	70 - 74	75 - 79	80 - 84	85 - 89
Number of people	1	12	7	5	3	2

Calculate: a. the mean

b. the median

c. the mode of the distribution

2. The data below is the load distribution in tones, a chain can support.

Load	83 - 85	86 - 88	89 - 91	92 - 94	95 - 97
Number of chain	2	8	5	14	1

Determine: a. the mean

b. the median

c. the mode of the weights

3. The data below is the ages, in years, of 50 people at a party.

Ages	1 - 20	21 - 40	41 - 60	61 - 80	81-100
Number of people	3	21	17	7	2

Determine: a. the mean

b. the median

c. the mode of the weights

CHAPTER 22
MEAN DEVIATION

The mean deviation of a set of data is the mean of the absolute deviation of the values from the mean of the group. The mean deviation for data not given in a frequency table is given by:

Mean deviation $= \dfrac{\sum |x - \bar{x}|}{N}$ where x is each value in the data, $\bar{x}$, is the mean and N is the number of values in the data.

For data given in a frequency table, the mean deviation is given by:

Mean deviation $= \dfrac{\sum f|x - \bar{x}|}{\sum f}$

Examples

1. Calculate the mean deviation of the following data: 2, 4, 1, 3, 0

Solution

Let us first calculate the mean of the data.

$$\text{Mean, } \bar{x} = \frac{2 + 4 + 1 + 3 + 0}{5}$$

$$= \frac{10}{5}$$

$$= 2$$

$\therefore \ \bar{x} = 2$

The deviation from the mean $(x - \bar{x})$ is now tabulated as follows.

Data (x)	$x - \bar{x}$ $(\bar{x} = 2)$	$|x - \bar{x}|$
2	0	0
4	2	2
1	-1	1
3	1	1
0	-2	2
		$\sum |x - \bar{x}| = 6$

$\therefore$ Mean deviation $= \dfrac{\sum |x - \bar{x}|}{N} = \dfrac{6}{5} = 1.2$

2. Calculate the mean deviation of the following data: 6, 2, 5, 8, 3, 6, 4, 5, 7, 4.

Solution

Let us first calculate the mean of the data.

$$\text{Mean, } \bar{X} = \frac{6+2+5+8+3+6+4+5+7+4}{10}$$

$$= \frac{50}{10}$$

$$= 5$$

$$\therefore \bar{x} = 5$$

The deviation from the mean $(x - \bar{x})$ is now tabulated as follows.

Data (x)	$x - \bar{x}$ $(\bar{x} = 5)$	$\lvert x - \bar{x} \rvert$
6	1	1
2	-3	3
5	0	0
8	3	3
3	-2	2
6	1	1
4	-1	1
5	0	0
7	2	2
4	-1	1
		$\sum \lvert x - \bar{x} \rvert = 14$

$$\therefore \quad \text{Mean deviation} = \frac{\sum \lvert x - \bar{x} \rvert}{N} = \frac{14}{10} = 1.4$$

3. The marks obtained by 40 students in a mathematics test are as shown below. Calculate the mean deviation of the data.

Marks	31 - 40	41 - 50	51 - 60	61 - 70	71 - 80	81 - 90	91- 100
Number of student	1	2	8	11	8	6	4

Solution

The table below summarizes the determination of the mean and the values needed for the mean deviation. Note that the mean used on the table has been calculated below the table.

Mark	mid-value x	$x - \bar{x}$ $\bar{x} = 69.75$	$\lvert x - \bar{x} \rvert$	No of student, f	fx	$f\lvert x - \bar{x} \rvert$
31 – 40	35.5	-34.25	34.25	1	35.5	34.25
41 – 50	45.5	-24.25	24.25	2	91	48.50
51 – 60	55.5	-14.24	14.24	8	444	114
61 – 70	65.5	-4.25	4.25	11	720.5	46.75
71 – 80	75.5	5.75	5.75	8	604	46
81 – 90	85.5	15.75	15.75	6	513	94.5
91 – 100	95.5	25.75	25.75	4	382	103
				$\sum f = 40$	$\sum fx = 2790$	

Mean, $\bar{x} = \dfrac{\sum fx}{\sum f} = \dfrac{2790}{40} = 69.75$

Using the values from the table above, $\sum f \lvert x - \bar{x} \rvert = 34.25 + 48.50 + 114 + 46.75 + 46 + 94.5 + 103 = 487$

$\therefore$ Mean deviation $= \dfrac{\sum f \lvert x - \bar{x} \rvert}{\sum f}$

$= \dfrac{487}{40}$

$\therefore$ Mean deviation $= 12.2$

4. The ages of 50 people in a hospital are as shown below. Calculate the mean deviation of the ages.

Age	1 - 5	6 - 10	11 – 15	16 - 20	21 - 25	26 - 30
Number of people	6	9	14	10	4	7

Solution

The table below summarizes the determination of the mean and the values needed for the mean deviation. The mean has been calculated below the table.

Age	mid-value x	$x - \bar{x}$ $\bar{x} = 14.8$	$\lvert x - \bar{x} \rvert$	No of people, f	fx	$f\lvert x - \bar{x} \rvert$
1 – 5	3	-11.8	11.8	6	18	70.8
6 – 10	8	-6.8	6.8	9	72	61.2
11 – 15	13	-1.8	1.8	14	182	25.2
16 – 20	18	3.2	3.2	10	180	32
21 – 25	23	8.2	8.2	4	92	32.8
26 – 30	28	13.2	13.2	7	196	92.4
				$\sum f = 50$	$\sum fx = 740$	

Mean, $\bar{x} = \dfrac{\sum fx}{\sum f} = \dfrac{740}{50} = 14.8$

Using the values from the table above, $\sum f|x - \bar{x}| = 70.8 + 61.2 + 25.2 + 32 + 32.8 + 92.4 = 314.4$

$\therefore$ Mean deviation $= \dfrac{\sum f|x - \bar{x}|}{\sum f}$

$\qquad\qquad = \dfrac{314.4}{50}$

$\therefore$ Mean deviation $= 6.29$

5. The table below shows the number of cars owned by some political public office holders.

Number of cars	1	2	3	4	5	6
Number of politicians	9	15	11	7	3	5

Calculate the mean deviation of the data.

Solution

The table below summarises the calculations of the mean and the mean deviation.

Cars x	$x - \bar{x}$ $\bar{x} = 2.9$	$|x - \bar{x}|$	No of politicians, f	Fx	$f|x - \bar{x}|$
1	-1.9	1.9	9	9	17.1
2	-0.9	0.9	15	30	13.5
3	0.1	0.1	11	33	1.1
4	1.1	1.1	7	28	7.7
5	2.1	2.1	3	15	6.3
6	3.1	3.1	5	30	15.5
			$\sum f = 50$	$\sum fx = 145$	

Mean, $\bar{x} = \dfrac{\sum fx}{\sum f} = \dfrac{145}{50} = 2.9$

Using the values from the table above, $\sum f|x - \bar{x}| = 17.1 + 13.5 + 1.1 + 7.7 + 6.3 + 15.5 = 61.2$

$\therefore$ Mean deviation $= \dfrac{\sum f|x - \bar{x}|}{\sum f}$

$\qquad\qquad = \dfrac{61.2}{50}$

$\therefore$ Mean deviation $= 1.224$

Exercise 22

1. Calculate the mean deviation of the following data: 0, 5, 7, 4, 5, 3

2. Calculate the mean deviation of the following data: 4, 6, 5, 9, 9, 5, 2, 4, 8, 6, 8

3. Calculate the mean deviation of the following data: 1, 3, 1, 4, 6

4. The marks obtained by 30 students in a physics test are as shown below. Calculate the mean deviation of the data.

Marks	0 - 9	10 - 19	20 - 29	30 - 39	40 - 49	50 – 59	60- 69
Number of student	4	1	5	8	3	2	7

5. The ages of 100 people in a village are as shown below. Calculate the mean deviation of the ages.

Age	11 - 20	21 - 30	31 – 40	41 - 50	51 - 60	61 - 70
Number of people	12	9	15	24	29	11

6. The number of employees in 50 enterprises are as shown below. Calculate the mean deviation of the data.

Marks	0 - 4	5 - 9	10 - 14	15 - 19	20 - 24	25 – 29	30- 34
Number of student	2	11	15	3	4	2	13

7. The table below shows the number of farms owned by some people in a city.

Number of farms	2	4	6	8	10	12
Number of people	3	5	10	6	8	8

Calculate the mean deviation of the data.

8. A die is rolled 50 times and the following data is obtained.

2	5	4	3	5	3	1	4	6	5	6	4	2
6	1	5	6	2	1	6	4	3	4	3	1	6
1	3	6	4	2	4	3	4	5	3	4	1	2
3	1	2	2	5	6	4	3	4	6	5		

a. Present the data in a frequency table

b. Calculate the mean deviation of the data.

CHAPTER 23
VARIANCE AND STANDARD DEVIATION

Variance is the mean of the squares of the deviations from the mean. Standard deviation is the positive square root of the variance.

Variance, standard deviation and mean deviation are also regarded as measures of dispersion or variation.

The variance of data not given on a frequency table is given by:

$$\text{Variance} = \frac{\Sigma(x - \bar{x})^2}{N}$$

For data given on a frequency table, the variance is given by:

$$\text{Variance} = \frac{\Sigma f(x - \bar{x})^2}{\Sigma f}$$

Standard deviation is the square root of variance.

Examples

1. Calculate the variance and standard deviation of the following data: 4, 2, 1, 5.

Solution

Let us first calculate the mean of the data.

$$\text{Mean, } \bar{x} = \frac{4 + 2 + 1 + 5}{4} = \frac{12}{4} = 3$$

We now present the deviation from the mean as follows.

Data x	$x - \bar{x}$ ($\bar{x} = 3$)	$(x - \bar{x})^2$
4	1	1
2	-1	1
1	-2	4
5	2	4
		$\Sigma(x - \bar{x})^2 = 10$

$$\text{Variance} = \frac{\Sigma(x - \bar{x})^2}{N} = = \frac{10}{4} = 2.5$$

∴ Standard deviation $= \sqrt{2.5} = 1.58$

2. Calculate the variance and standard deviation of the data below:
   2, 5, 3, 2, 6, 5, 7, 2.

Solution

Let us first calculate the mean of the data as follows:

$$\text{Mean, } \bar{x} = \frac{2 + 5 + 3 + 2 + 6 + 5 + 7 + 2}{8}$$

$$= \frac{32}{8}$$

$$= 4$$

The deviation from the mean is as presented below.

Data x	$x - \bar{x}$ $(\bar{x} = 4)$	$(x - \bar{x})^2$
2	-2	4
5	1	1
3	-1	1
2	-2	4
6	2	4
5	1	1
7	3	9
2	-2	4
		$\sum(x - \bar{x})^2 = 28$

$$\text{Variance} = \frac{\sum(x - \bar{x})^2}{N} = \frac{28}{8} = 3.5$$

$\therefore$ Standard deviation $= \sqrt{3.5} = 1.87$

3. The distances in Km, from school to the homes of 30 students are as shown below. Calculate:

a. the variance

b. the standard deviation of the data

Distance (Km)	0 – 4	5 - 9	10 – 14	15 - 19	20 - 24	25 - 29
Number of students	2	10	8	6	3	1

Solutions

The working is set out as shown on the table below

Distance	mid-value x	No of student f	fx	$x - \bar{x}$ $\bar{x} = 12.2$	$(x - \bar{x})^2$	$f(x - \bar{x})^2$
0 – 4	2	2	4	-10.2	104.04	208.08
5 – 9	7	10	70	-5.2	27.04	270.4
10 – 14	12	8	96	-0.2	0.04	0.32
15 – 19	17	6	102	4.8	23.04	138.24
20 – 24	22	3	66	9.8	96.04	288.12
25 – 29	27	1	27	14.8	219.04	219.04
		$\sum f = 30$	$\sum fx = 365$			

Mean, $\bar{x} = \dfrac{\sum fx}{\sum f} = \dfrac{365}{30} = 12.2$

a. Using the values from the table above, $\sum f(x - \bar{x})^2 = 208.08 + 270.4 + 0.32 + 138.24 + 288.12 + 219.04 = 1124.2$

∴   Variance $= \dfrac{\sum f(x - \bar{x})^2}{\sum f}$

$= \dfrac{1124.2}{30}$

∴   Variance = 37.5

b. Standard deviation $= \sqrt{\text{Variance}}$

$= \sqrt{37.5}$

∴   Standard deviation = 6.1

4. The projected population in millions, of 20 states in a country are as shown below. Calculate:

a. the variance

b. the standard deviation of the data

Population	1 - 5	6 - 10	11 – 15	16 - 20	21 - 25	26 - 30
Number of state	1	8	5	3	2	1

Solutions

The working is set out as shown on the table below

Popula-tion	mid-value x	No of states f	fx	$x - \bar{x}$ $\bar{x} = 13$	$(x - \bar{x})^2$	$f(x - \bar{x})^2$
1 – 5	3	1	3	-10	100	100
6 – 10	8	8	64	-5	25	200
11 – 15	13	5	65	0	0	0
16 – 20	18	3	54	5	25	75
21 – 25	23	2	46	10	100	200
26 – 30	28	1	28	15	225	225
		$\Sigma f = 20$	$\Sigma fx = 260$			

Mean, $\bar{x} = \dfrac{\Sigma fx}{\Sigma f} = \dfrac{260}{20} = 13$

a. Using the values from the table above, $\Sigma f(x - \bar{x})^2 = 100 + 200 + 0 + 75 + 200 + 225 = 800$

$\therefore \quad$ Variance $= \dfrac{\Sigma f(x - \bar{x})^2}{\Sigma f}$

$\qquad = \dfrac{800}{20}$

$\therefore \quad$ Variance = 40

b. Standard deviation $= \sqrt{\text{Variance}}$

$\qquad\qquad\qquad = \sqrt{40}$

$\therefore$ Standard deviation = 6.3

5. The scores obtained by 100 students in a test are as shown below. Calculate:
a. the variance
b. the standard deviation of the scores

Scores	2	3	4	5	6	7
Number of student	10	22	18	30	12	8

Solutions
The working is set out as shown on the table below.

Scores x	No of students f	fx	$x - \bar{x}$ $\bar{x} = 4.4$	$(x - \bar{x})^2$	$f(x - \bar{x})^2$
2	12	24	-2.4	5.76	69.12
3	18	54	-1.4	1.96	35.28
4	22	88	-0.4	0.16	3.52
5	24	120	0.6	0.36	8.64
6	14	84	1.6	2.56	35.84
7	10	70	2.6	6.76	67.6
	$\Sigma f = 100$	$\Sigma fx = 440$			

Mean, $\bar{x} = \dfrac{\Sigma fx}{\Sigma f} = \dfrac{440}{100} = 4.4$

a. Using the values from the table above, $\Sigma f(x - \bar{x})^2 = 69.12 + 35.28 + 3.52 + 8.64 + 35.84 + 67.6$
= 220

$\therefore$ Variance $= \dfrac{\Sigma f(x - \bar{x})^2}{\Sigma f}$

$= \dfrac{220}{100}$

$\therefore$ Variance = 2.2

b. Standard deviation $= \sqrt{\text{Variance}}$

$= \sqrt{2.2}$

$\therefore$ Standard deviation = 1.48

Exercise 23

1. Calculate the variance and standard deviation of the following data: 3, 5, 4, 7, 6.

2. Calculate the variance and standard deviation of the data below:

    1, 0, 4, 3, 5, 8, 6, 4, 7, 2.

3. The scores of 50 students in a test are as shown below. Calculate:
a. the variance
b. the standard deviation of the data

Scores	0 – 9	10 - 19	20 – 29	30 - 39	40 – 49	50 - 59
Number of students	5	12	6	18	5	4

4. The ages of employees in an organization are as shown below. Calculate:

a. the variance

b. the standard deviation of the data

Age	20 - 24	25 - 29	30 - 34	35 - 39	40 - 44
Number of empoyees	8	6	3	1	2

5. The scores obtained by 40 students in a test are as shown below. Calculate:

a. the variance

b. the standard deviation of the scores

Scores	5	6	7	8	9	10
Number of student	1	2	4	12	20	1

CHAPTER 24
QUARTILES AND PERCENTILES BY INTERPOLATION METHOD

When a distribution is divide into four equal parts, it is called a quartile. When it is divided into hundred equal parts, such a division is called percentile.

The first quartile is also called lower quartile, and it is denoted by Q_1.

The second quartile is also called median, and it is denoted by Q_2.

The third quartile is also called upper quartile, and it is denoted by Q_3.

The lower quartile for a grouped data is calculated as follows:

$$Q_1 = L_1 + C\left(\frac{\frac{\Sigma f}{4} - CF_{bQ_1}}{F_{Q_1}}\right)$$

Where, $\frac{\Sigma f}{4}$ determines the lower quartile class

L_1 = Lower class boundary of the lower quartile class

CF_{bQ_1} = Cumulative frequency before the lower quartile class

F_{Q_1} = Frequency of the lower quartile class

C = Class width

The median is calculated as follows:

$$Q_2 = L_2 + C\left(\frac{\frac{\Sigma f}{2} - CF_{bm}}{F_m}\right)$$

Where, $\frac{\Sigma f}{2}$ determines the median class

L_2 = Lower class boundary of the median class

CF_{bm} = Cumulative frequency before the median class

F_m = Frequency of the median class

C = Class width

The upper quartile is calculated as follows:

$$Q_3 = L_3 + C\left(\frac{\frac{3\Sigma f}{4} - CF_{bQ_3}}{F_{Q_3}}\right)$$

Where, $\frac{3\Sigma f}{4}$ determines the upper quartile class

L_3 = Lower class boundary of the upper quartile class

CF_{bQ_3} = Cumulative frequency before the upper quartile class

F_{Q_3} = Frequency of the upper quartile class

C = Class width

The interquartile range is given by:

Interquartile range = $Q_3 - Q_1$

The semi-interquartile range is also called quartile deviation, and it is given by:

Semi-interquartile range = $\dfrac{Q_3 - Q_1}{2}$

The percentile is calculated as follows:

$$P_N = L_N + C\left(\dfrac{\frac{N \Sigma f}{100} - CF_{bP_N}}{F_{P_N}}\right)$$

Where P_N is the N percentile, and $\dfrac{N \Sigma f}{100}$ determines the N percentile class

L_N = Lower class boundary of the N percentile class

CF_{bP_N} = Cumulative frequency before the N percentile class

F_{P_N} = Frequency of the N percentile class

C = Class width

Examples

1. The following is the record of marks of 40 students in an examination:

64 84 91 58 43 86 73 33 76 80 57 33 53 29 40 27 72 19 51
67 37 14 18 92 13 45 61 39 23 22 22 41 27 51 63 47 19 35
39 76

Using class interval 11 – 20, 21 – 30, …, prepare a frequency table for the distribution.
Hence calculate the:

a. median

b. lower quartile

c. upper quartile

d. interquartile range

e. quartile deviation/semi-interquartile range

f. 40th percentile

g. 85th percentile

Solutions

The frequency table is as shown below.

Class interval	Frequency
11 – 20	5
21 – 30	6
31 – 40	7
41 – 50	4
51 – 60	5
61 – 70	4
71 – 80	5
81 – 90	2
91 – 100	2

a. In order to calculate the median, a table of the class boundaries and cumulative frequency has to be drawn as shown below.

Class interval	Class boundary	Frequency	Cumulative frequency	Class width
11 – 20	10.5 – 20.5	5	5	10
21 – 30	20.5 – 30.5	6	11	10
31 – 40	30.5 – 40.5	7	18	10
41 – 50	40.5 – 50.5	4	22	10
51 – 60	50.5 – 60.5	5	27	10
61 – 70	60.5 – 70.5	4	31	10
71 – 80	70.5 – 80.5	5	36	10
81 – 90	80.5 – 90.5	2	38	10
91 - 100	90.5 – 100.5	2	40	10

The median is calculated as follows:

$$Q_2 = L_2 + C\left(\frac{\frac{\Sigma f}{2} - CF_{bm}}{F_m}\right)$$

$\frac{\Sigma f}{2} = \frac{40}{2} = 20$. This shows that the median class is at the 20th position. This is the class, 41 – 50. This position is obtained by counting the frequency to get to the 20th position. 5 + 6 + 7 = 18. This shows that the 18th position is occupied by the class 31 – 40. After this class, the next frequency is 4. When 4 positions are added to 18 positions, it gives 22. This means that these 4 positions are the 19th, 20th, 21st and 22nd positions. These 4 positions are occupied by the class 41 – 50 as shown on the table. Hence the class in the 20th position is 41 – 50. You can also look at the cumulative frequency and see where the 20th position class falls.

L_2 = Lower class boundary of the median class = 40.5

CF_{bm} = Cumulative frequency before the median class = 18

F_m = Frequency of the median class = 4

C = Class width = 10. The class limit is the difference between an upper class boundary and a lower class boundary. For example, 20.5 − 10.5 = 10.

$$\therefore \quad Q_2 = L_2 + C\left(\frac{\frac{\Sigma f}{2} - CF_{bm}}{F_m}\right)$$

$$= 40.5 + 10\left(\frac{\frac{40}{2} - 18}{4}\right)$$

$$= 40.5 + 10\left(\frac{20 - 18}{4}\right)$$

$$= 40.5 + 10\left(\frac{2}{4}\right)$$

$$= 40.5 + \left(\frac{10 \times 2}{4}\right)$$

$$= 40.5 + 5$$

$$Q_2 = 45.5$$

b. The lower quartile is calculated as follows:

$$Q_1 = L_1 + C\left(\frac{\frac{\Sigma f}{4} - CF_{bQ_1}}{F_{Q_1}}\right)$$

$\frac{\Sigma f}{4} = \frac{40}{4} = 10$. Hence the lower quartile class is at the 10th position. This class is, 21 − 30.

L_1 = Lower class boundary of the lower quartile class = 20.5

CF_{bQ_1} = Cumulative frequency before the lower quartile class = 5

F_{Q_1} = Frequency of the lower quartile class = 6

C = Class width = 10

$$\therefore \quad Q_1 = L_1 + C\left(\frac{\frac{\Sigma f}{4} - CF_{bQ_1}}{F_{Q_1}}\right)$$

$$= 20.5 + 10\left(\frac{\frac{40}{4} - 5}{6}\right)$$

$$= 20.5 + 10\left(\frac{10 - 5}{6}\right)$$

$$= 20.5 + 10\left(\frac{5}{6}\right)$$

$$= 20.5 + \left(\frac{10 \times 5}{6}\right)$$

$$= 20.5 + 8.3$$

$$Q_1 = 28.8$$

c. The upper quartile is calculated as follows:

$$Q_3 = L_3 + C\left(\frac{\frac{3\Sigma f}{4} - CF_{bQ_3}}{F_{Q_3}}\right)$$

$\frac{3\Sigma f}{4} = \frac{3 \times 40}{4} = 30$. Hence the upper quartile class is at the 30th position. This class is, 61 − 70.

L_3 = Lower class boundary of the upper quartile class = 60.5

CF_{bQ_3} = Cumulative frequency before the upper quartile class = 27

F_{Q_3} = Frequency of the upper quartile class = 4

C = Class width = 10

$$\therefore \quad Q_3 = L_3 + C\left(\frac{\frac{3\Sigma f}{4} - CF_{bQ_3}}{F_{Q_3}}\right)$$

$$= 60.5 + 10\left(\frac{\frac{3 \times 40}{4} - 27}{4}\right)$$

$$= 60.5 + 10\left(\frac{30 - 27}{4}\right)$$

$$= 60.5 + 10\left(\frac{3}{4}\right)$$

$$= 60.5 + \left(\frac{10 \times 3}{4}\right)$$

$$= 60.5 + 7.5$$

$$Q_3 = 68$$

d. Interquartile range = $Q_3 - Q_1$

$$= 68 - 28.8$$

$$= 39.2$$

e. Quartile deviation/semi-interquartile range is given by:

$$Q = \frac{Q_3 - Q_1}{2}$$

$$= \frac{68 - 28.8}{2}$$

$$= \frac{39.2}{2}$$

$$Q = 19.6$$

f. The 40th percentile is calculated as follows:

$$P_N = L_N + C\left(\frac{\frac{N\Sigma f}{100} - CF_{bP_N}}{F_{P_N}}\right)$$

$$P_N = P_{40}$$

$\frac{N\Sigma f}{100} = \frac{40 \times 40}{100}$ = 16. Hence the 40th percentile class is at the 16th position. This class is: 31 - 40

$L_N = L_{40}$ = Lower class boundary of the 40th percentile class = 30.5

$CF_{bP_N} = CF_{bP_{40}}$ = Cumulative frequency before the 40th percentile class = 11

$F_{P_N} = F_{P_{40}}$ = Frequency of the 40th percentile class = 7

C = Class width = 10

Hence, $P_{40} = L_{40} + C\left(\dfrac{\frac{40\sum f}{100} - CF_{bP_{40}}}{F_{P_{40}}}\right)$

$= 30.5 + 10\left(\dfrac{\frac{40 \times 40}{100} - 11}{7}\right)$

$= 30.5 + 10\left(\dfrac{16 - 11}{7}\right)$

$= 30.5 + 10\left(\dfrac{5}{7}\right)$

$= 30.5 + \left(\dfrac{10 \times 5}{7}\right)$

$= 30.5 + 7.1$

$P_{40} = 37.6$

g. The 85th percentile is calculated as follows:

$P_N = L_N + C\left(\dfrac{\frac{N\sum f}{100} - CF_{bP_N}}{F_{P_N}}\right)$

$P_N = P_{85}$

$\dfrac{N\sum f}{100} = \dfrac{85 \times 40}{100} = 34$. Hence the 85th percentile class is at the 34th position. This class is: 71 - 80

$L_N = L_{85}$ = Lower class boundary of the 85th percentile class = 70.5

$CF_{bP_N} = CF_{bP_{85}}$ = Cumulative frequency before the 85th percentile class = 31

$F_{P_N} = F_{P_{85}}$ = Frequency of the 85th percentile class = 5

C = Class width = 10

Hence, $P_{85} = L_{85} + C\left(\dfrac{\frac{85 \times 40}{100} - CF_{bP_{85}}}{F_{P_{85}}}\right)$

$= 70.5 + 10\left(\dfrac{\frac{85 \times 40}{100} - 31}{5}\right)$

$= 70.5 + 10\left(\dfrac{34 - 31}{5}\right)$

$= 70.5 + 10\left(\dfrac{3}{5}\right)$

$= 70.5 + \left(\dfrac{10 \times 3}{5}\right)$

$= 70.5 + 6$

$P_{85} = 76.5$

2. The table below shows the distribution of marks scored by students in an examination.

Class interval	Frequency
60 – 64	2
65 – 69	4
70 – 74	7
75 – 79	13
80 – 84	10
85 – 89	8
90 – 94	5
95 – 99	1

From the data, calculate:

a. median

b. lower quartile

c. upper quartile

d. interquartile range

e. semi-interquartile range

f. 70th percentile

g. the pass mark if 25% of the students passed

h. the pass mark if it was later agreed that only 40% of the students should fail.

Solution

a. In order to calculate the median, a table of the class boundaries and cumulative frequency has to be drawn as shown below.

Class interval	Class boundary	Frequency	Cumulative frequency	Class width
60 – 64	59.5 – 64.5	2	2	5
65 – 69	64.5 – 69.5	4	6	5
70 – 74	69.5 – 74.5	7	13	5
75 – 79	74.5 – 79.5	13	26	5
80 – 84	79.5 – 84.5	10	36	5
85 – 89	84.5 – 89.5	8	44	5
90 – 94	89.5 – 94.5	5	49	5
95 – 99	94.5 – 99.5	1	50	5

The median is calculated as follows:

$$Q_2 = L_2 + C\left(\frac{\frac{\Sigma f}{2} - CF_{bm}}{F_m}\right)$$

$\frac{\Sigma f}{2} = \frac{50}{2} = 25$. This shows that the median class is at the 25th position. This is the class, 75 – 79.

This is obtained by looking at the cumulative frequency to see where the 25th position class falls.

L_2 = Lower class boundary of the median class = 74.5

CF_{bm} = Cumulative frequency before the median class = 13

F_m = Frequency of the median class. This is also 13

C = Class width = 5

$$\therefore \quad Q_2 = L_2 + C\left(\frac{\frac{\Sigma f}{2} - CF_{bm}}{F_m}\right)$$

$$= 74.5 + 5\left(\frac{\frac{50}{2} - 13}{13}\right)$$

$$= 74.5 + 5\left(\frac{25 - 13}{13}\right)$$

$$= 74.5 + 5\left(\frac{12}{13}\right)$$

$$= 74.5 + \left(\frac{5 \times 12}{13}\right)$$

$$= 74.5 + 4.6$$

$$Q_2 = 79.1$$

b. The lower quartile is calculated as follows:

$$Q_1 = L_1 + C\left(\frac{\frac{\Sigma f}{4} - CF_{bQ_1}}{F_{Q_1}}\right)$$

$\frac{\Sigma f}{4} = \frac{50}{4}$ = 12.5. Hence the lower quartile class is at the 12th or 13th position. This class is, 70 – 74.

L_1 = Lower class boundary of the lower quartile class = 69.5

CF_{bQ_1} = Cumulative frequency before the lower quartile class = 6

F_{Q_1} = Frequency of the lower quartile class = 7

C = Class width = 5

$$\therefore \quad Q_1 = L_1 + C\left(\frac{\frac{\Sigma f}{4} - CF_{bQ_1}}{F_{Q_1}}\right)$$

$$= 69.5 + 5\left(\frac{\frac{50}{4} - 6}{7}\right)$$

$$= 69.5 + 5\left(\frac{12.5 - 6}{7}\right)$$

$$= 69.5 + 5\left(\frac{6.5}{7}\right)$$

$$= 69.5 + \left(\frac{5 \times 6.5}{7}\right)$$

$$= 69.5 + 4.6$$

$$Q_1 = 74.1$$

c. The upper quartile is calculated as follows:

$$Q_3 = L_3 + C\left(\frac{\frac{3\sum f}{4} - CF_{bQ3}}{F_{Q3}}\right)$$

$\frac{3\sum f}{4} = \frac{3 \times 50}{4} = 37.5$. Hence the upper quartile class is at the 37th or 38th position. This class is, 85 – 89.

L_3 = Lower class boundary of the upper quartile class = 84.5

CF_{bQ_3} = Cumulative frequency before the upper quartile class = 36

F_{Q_3} = Frequency of the upper quartile class = 8

C = Class width = 5

$$\therefore \quad Q_3 = L_3 + C\left(\frac{\frac{3\sum f}{4} - CF_{bQ3}}{F_{Q3}}\right)$$

$$= 84.5 + 5\left(\frac{\frac{3 \times 50}{4} - 36}{8}\right)$$

$$= 84.5 + 5\left(\frac{37.5 - 36}{8}\right)$$

$$= 84.5 + 5\left(\frac{1.5}{8}\right)$$

$$= 84.5 + 0.9$$

$$Q_3 = 85.4$$

d. Interquartile range = $Q_3 - Q_1$

$$= 85.4 - 74.1$$

$$= 11.3$$

e. Semi-interquartile range, $Q = \frac{Q_3 - Q_1}{2}$

$$= \frac{85.4 - 74.1}{2}$$

$$= \frac{11.3}{2}$$

$$Q = 5.65$$

f. The 70th percentile is calculated as follows:

$$P_N = L_N + C\left(\frac{\frac{N\sum f}{100} - CF_{bPN}}{F_{PN}}\right)$$

$P_N = P_{70}$

$\frac{N\sum f}{100} = \frac{70 \times 50}{100} = 35$. Hence the 70th percentile class is at the 35th position. This class is: 80 - 84

$L_N = L_{70}$ = Lower class boundary of the 70th percentile class = 79.5

$CF_{bPN} = CF_{bP_{70}}$ = Cumulative frequency before the 70th percentile class = 26

$F_{P_N} = F_{P_{70}}$ = Frequency of the 70th percentile class = 10

C = Class width = 5

Hence, $P_{70} = L_{70} + C\left(\dfrac{\frac{70\sum f}{100} - CF_{bP_{70}}}{F_{P_{70}}}\right)$

$= 79.5 + 5\left(\dfrac{\frac{70 \times 50}{100} - 26}{10}\right)$

$= 79.5 + 5\left(\dfrac{35 - 26}{10}\right)$

$= 79.5 + 5\left(\dfrac{9}{10}\right)$

$= 79.5 + 4.5$

$P_{70} = 84$

g. If 25% of the students passed, then the first 75% (i.e. 100 – 25) of the students failed. This means that the pass mark is at the 75th percentile.

Note that the pass mark is always at the failure percentile.

Hence the 75th percentile is calculated as follows:

$P_N = L_N + C\left(\dfrac{\frac{N\sum f}{100} - CF_{bP_N}}{F_{P_N}}\right)$

$P_N = P_{75}$

$\dfrac{N\sum f}{100} = \dfrac{75 \times 50}{100}$ = 37.5. Hence the 75th percentile class is at the 37.5th position. This class is: 85 - 89

$L_N = L_{75}$ = Lower class boundary of the 75th percentile class = 84.5

$CF_{bP_N} = CF_{bP_{75}}$ = Cumulative frequency before the 75th percentile class = 36

$F_{P_N} = F_{P_{75}}$ = Frequency of the 75th percentile class = 8

C = Class width = 5

Hence, $P_{75} = L_{75} + C\left(\dfrac{\frac{75\sum f}{100} - CF_{bP_{75}}}{F_{P_{75}}}\right)$

$= 84.5 + 5\left(\dfrac{\frac{75 \times 50}{100} - 36}{8}\right)$

$= 84.5 + 5\left(\dfrac{37.5 - 36}{8}\right)$

$= 84.5 + 5\left(\dfrac{1.5}{8}\right)$

$= 84.5 + 0.9$

$P_{75} = 85.4$

Hence the pass mark is 85.4

h. If 40% of the students should fail, then the pass mark is at the 40th percentile. Hence the 40th percentile is calculated as follows:

$$P_N = L_N + C\left(\frac{\frac{N\Sigma f}{100} - CF_{bP_N}}{F_{P_N}}\right)$$

$P_N = P_{40}$

$\frac{N\Sigma f}{100} = \frac{40 \times 50}{100} = 20$. Hence the 40th percentile class is at the 20th position. This class is: 75 - 79

$L_N = L_{40}$ = Lower class boundary of the 40th percentile class = 74.5

$CF_{bP_N} = CF_{bP_{40}}$ = Cumulative frequency before the 40th percentile class = 13

$F_{P_N} = F_{P_{40}}$ = Frequency of the 40th percentile class = 13

C = Class width = 5

Hence, $P_{40} = L_{40} + C\left(\frac{\frac{40\Sigma f}{100} - CF_{bP_{40}}}{F_{P_{40}}}\right)$

$$= 74.5 + 5\left(\frac{\frac{40 \times 50}{100} - 13}{13}\right)$$

$$= 74.5 + 5\left(\frac{20 - 13}{13}\right)$$

$$= 74.5 + 5\left(\frac{7}{13}\right)$$

$$= 74.5 + 2.7$$

$P_{40} = 77.2$

Hence the pass mark is 77.2

3. The table below shows the masses of some items sold in a supermarket.

Mass	1.5 − 1.9	2.0 − 2.4	2.5 − 2.9	3.0 − 3.4	3.5 − 3.9	4.0 − 4.5
Number of Items	5	12	6	18	5	4

From the table given above, estimate:

a. median

b. lower quartile

c. upper quartile

d. the 55th percentile

Solution

a. In order to calculate the median, a table of the class boundaries and cumulative frequency has to be drawn as shown below.

Class interval	Class boundary	Frequency	Cumulative frequency	Class width
1.5 – 1.9	1.45 – 1.95	5	5	0.5
2.0 – 2.4	1.95 – 2.45	12	17	0.5
2.5 – 2.9	2.45 – 2.95	6	23	0.5
3.0 – 3.4	2.95 – 3.45	18	41	0.5
3.5 – 3.9	3.45 – 3.95	5	46	0.5
4.0 – 4.4	3.95 – 4.45	4	50	0.5

Note that in computing the class boundaries, a difference between an upper class limit and a lower class limit, such as, $2.0 - 1.9 = 0.1$, is first determined and then divided by 2 to give, $0.1/2 = 0.05$. It is this 0.05 that is added and subtracted from the class limit values to obtain the class boundary values. This is the method applied in obtaining the class boundaries of any given grouped data.

The median is calculated as follows:

$$Q_2 = L_2 + C\left(\frac{\frac{\Sigma f}{2} - CF_{bm}}{F_m}\right)$$

$\frac{\Sigma f}{2} = \frac{50}{2} = 25$. Hence the median class is at the 25th position. This is the class, $3.0 - 3.4$. This is obtained by looking at the cumulative frequency to see where the 25th position class falls.

L_2 = Lower class boundary of the median class = 2.95

CF_{bm} = Cumulative frequency before the median class = 23

F_m = Frequency of the median class = 18

C = Class width = 0.5, i.e. $1.95 - 1.45 = 0.5$.

$$\therefore \quad Q_2 = L_2 + C\left(\frac{\frac{\Sigma f}{2} - CF_{bm}}{F_m}\right)$$

$$= 2.95 + 0.5\left(\frac{\frac{50}{2} - 23}{18}\right)$$

$$= 2.95 + 0.5\left(\frac{25 - 23}{18}\right)$$

$$= 2.95 + 0.5\left(\frac{2}{18}\right)$$

$$= 2.95 + 0.06$$

$$Q_2 = 3.01$$

b. The lower quartile is calculated as follows:

$$Q_1 = L_1 + C\left(\dfrac{\frac{\Sigma f}{4} - CF_{bQ1}}{F_{Q1}}\right)$$

$\dfrac{\Sigma f}{4} = \dfrac{50}{4} = 12.5$. This shows that the lower quartile class is at the 12th or 13th position. This class is, 2.0 – 2.4.

L_1 = Lower class boundary of the lower quartile class = 1.95

CF_{bQ_1} = Cumulative frequency before the lower quartile class = 5

F_{Q_1} = Frequency of the lower quartile class = 12

C = Class width = 0.5

$$\therefore \quad Q_1 = L_1 + C\left(\dfrac{\frac{\Sigma f}{4} - CF_{bQ1}}{F_{Q1}}\right)$$

$$= 1.95 + 0.5\left(\dfrac{\frac{50}{4} - 5}{12}\right)$$

$$= 1.95 + 0.5\left(\dfrac{12.5 - 5}{12}\right)$$

$$= 1.95 + 0.5\left(\dfrac{7.5}{12}\right)$$

$$= 1.95 + 0.31$$

$$Q_1 = 2.26$$

c. The upper quartile is calculated as follows:

$$Q_3 = L_3 + C\left(\dfrac{\frac{3\Sigma f}{4} - CF_{bQ3}}{F_{Q3}}\right)$$

$\dfrac{3\Sigma f}{4} = \dfrac{3 \times 50}{4} = 37.5$. This shows that the upper quartile class is at the 37th or 38th position. This class is, 3.0 – 3.4.

L_3 = Lower class boundary of the upper quartile class = 2.95

CF_{bQ_3} = Cumulative frequency before the upper quartile class = 23

F_{Q_3} = Frequency of the upper quartile class = 18

C = Class width = 0.5

$$\therefore \quad Q_3 = L_3 + C\left(\dfrac{\frac{3\Sigma f}{4} - CF_{bQ3}}{F_{Q3}}\right)$$

$$= 2.95 + 0.5\left(\dfrac{\frac{3 \times 50}{4} - 23}{18}\right)$$

$$= 2.95 + 0.5\left(\dfrac{37.5 - 23}{18}\right)$$

$$= 2.95 + 0.5\left(\dfrac{14.5}{18}\right)$$

$$= 2.95 + 0.4$$

$$Q_3 = 3.35$$

d. The 55th percentile is calculated as follows:

$$P_N = L_N + C\left(\frac{\frac{N\Sigma f}{100} - CF_{bP_N}}{F_{P_N}}\right)$$

$$P_N = P_{55}$$

$$\frac{N\Sigma f}{100} = \frac{55 \times 50}{100} = 27.5.$$ Hence the 55th percentile class is at the 27th and 28th position. This class is: 3.0 – 3.4

$L_N = L_{55}$ = Lower class boundary of the 55th percentile class = 2.95

$CF_{bP_N} = CF_{bP_{55}}$ = Cumulative frequency before the 55th percentile class = 23

$F_{P_N} = F_{P_{55}}$ = Frequency of the 55th percentile class = 18

C = Class width = 0.5

Hence, $P_{55} = L_{55} + C\left(\dfrac{\frac{55\Sigma f}{100} - CF_{bP_{55}}}{F_{P_{55}}}\right)$

$$= 2.95 + 0.5\left(\frac{\frac{55 \times 50}{100} - 23}{18}\right)$$

$$= 2.95 + 0.5\left(\frac{27.5 - 23}{18}\right)$$

$$= 2.95 + 0.5\left(\frac{4.5}{18}\right)$$

$$= 2.95 + 0.13$$

$$P_{55} = 3.08$$

Exercise 24

1. The following is the record of marks of 40 students in an examination:

34 74 92 58 46 76 73 23 66 70 57 43 53 39

50 37 82 29 54 77 67 19 18 96 15 55 41 29

33 52 22 81 77 81 58 27 20 55 49 96

Using class interval 11 – 20, 21 – 30, …, prepare a frequency table for the distribution. Hence calculate the:

a. median

b. lower quartile

c. upper quartile

d. interquartile range

e. quartile deviation/semi-interquartile range

f. 30th percentile

g. 68th percentile

2. The table below shows the distribution of marks scored by students in an examination.

Class interval	Frequency
10 – 14	1
15 – 19	3
20 – 24	8
25 – 29	11
30 – 34	7
35 – 39	9
40 – 44	10
45 – 49	5

From the data, calculate:

a. median

b. lower quartile

c. upper quartile

d. interquartile range

e. semi-interquartile range

f. 80th percentile

g. the pass mark if 35% of the students passed

h. the pass mark if 15% the students should fail.

3. The table below shows the height of some flowers sold in a farm.

Mass	0.5 – 0.9	1.0 – 1.4	1.5 – 1.9	2.0 – 2.4	2.5 – 2.9	3.0 – 3.4
No of Items	4	15	12	9	7	3

From the table given above, estimate:

a. median

b. lower quartile

c. upper quartile

d. the 45th percentile

e. pass mark if 90% of the students passed

4. The table below shows the distribution of marks scored by students in an test.

Class interval	Frequency
0 – 4	1
5 – 9	4
10 – 14	7
15 – 19	5
20 – 24	9
25 – 29	1
30 – 34	2
35 – 39	1

From the data, calculate:

a. median

b. lower quartile

c. upper quartile

d. interquartile range

e. semi-interquartile range

f. 60th percentile

g. the pass mark if 10% of the students passed

5. The table below shows the weight in gram of some seeds found in some cocoa pods.

Mass	0 – 0.4	0.5 – 0.9	1.0 – 1.4	1.5 – 1.9	2.0 – 2.4	2.5 – 2.9
No of Items	9	21	16	22	28	4

From the table given above, estimate:

a. median

b. lower quartile

c. upper quartile

d. the 25th percentile

e. pass mark if 68% of the students passed

CHAPTER 25
THE BASIC THEORY OF PROBABILITY

Probability is the likelihood of an event happening. Mathematically probability is given by:

$$\text{Probability} = \frac{\text{number of required outcome}}{\text{number of total or possible outcome}}$$

If the probability of an event happening is x, then the probability that it will not happen will be given by: $1 - x$

Probability must lie between the values of 0 and 1. If an event cannot happen, then its probably is 0. If an event is certain to happen, then its probability is 1.

Mutually Exclusive Events

When there is no member/element common between two or more similar events, then we say they are mutually exclusive events. For example the event of odd numbers or even numbers are mutually exclusive. They are disjoint sets.

Addition Law of Probability

If two events are mutually exclusive, then the probability of one or the other happening is the sum of their individual probabilities.

Independent Events

When a die is thrown, and a coin is tossed, these two events have no effect on each other. Such events are called independent events

Product law of probability

If two events are independent, then the probability of both events happening is the product is the product (multiplication) of their individual probabilities.

CHAPTER 26
PROBABILITY ON SIMPLE EVENTS

Examples

1. The table below give the number of students in each age group in a class.

Age (Years)	12	13	14	15	16	17
number of students	6	3	10	4	2	5

If a student is chosen at random, find the probability that the student is:

(a) 13 years old

(b) 15 years old or less

(c) at least 16 years old

(d) most 13 years old

(e) not 17 years old

<u>Solution</u>

(a) Pr. (13 years old) $= \dfrac{\text{Number of students who are 13 years old}}{\text{Total number of students}}$

$= \dfrac{3}{30}$

$= \dfrac{1}{10}$ (when $\dfrac{3}{30}$ is express in its lowest term, it gives $\dfrac{1}{10}$)

(b) Pr. (15 years or less) $= \dfrac{\text{Students who are 15 years and below}}{\text{Total number of students}}$

$= \dfrac{4+10+3+6}{30}$

$= \dfrac{23}{30}$

(c) Pr. (At least 16 years old) $= \dfrac{\text{Students who are 16 years and above}}{\text{Total number of students}}$

$= \dfrac{2+5}{30}$

$= \dfrac{7}{30}$

(d) Pr. (At most 13 years) $= \dfrac{\text{Students who are 13 years and below}}{\text{Total number of students}}$

$= \dfrac{3+6}{30}$

$= \dfrac{9}{30}$

$= \dfrac{3}{10}$ (When expressed in its lowest term)

(e) Pr. (17 years old) $= \dfrac{\text{number of students who are 17 years old}}{\text{total number of students}}$

$= \dfrac{5}{30}$

$= \dfrac{1}{6}$

Therefore, Pr. (Not 17 years old) = 1 - Pr. (17 years old)

$= 1 - \dfrac{1}{6}$

$= \dfrac{5}{6}$

2. The probability that a seed will germinate is $\dfrac{2}{5}$. What is the probability that it will not germinate?

Solution

Pr. (It will germinate) $= \dfrac{2}{5}$

Pr. (It will not germinate) $= 1 - \dfrac{3}{5}$

$= \dfrac{2}{5}$

3. A letter is chosen at random from the alphabet. Find the probability that it is one of the letters of the word: PROBABILITY.

Solution

In this case a letter should not be counted more than once. Avoiding repetition, the word can now be written as:

PROBALITY (i.e. 9 letters). Note that there are 26 letters of the alphabet.

Therefore, Pr. (one letter from PROBABILITY) $= \dfrac{9}{26}$

4. The probability that a boy gains admission into a higher institution is $\dfrac{3}{7}$. What is the probability that he does not gain admission into the institution?

Solution

Pr. (He gains admission) $= \dfrac{3}{7}$

Pr. (He does not gain admissions) $= 1 - \dfrac{3}{7}$

$$= \dfrac{4}{7}$$

5. Out of every 100 cars, 4 develop mechanical fault within 6 months of purchase. What is the probability of buying a car which will not develop a mechanical fault within 6 months of purchase?

Solution

Total number of cars is 100. Number of cars with fault within 6 months is 4. Number of cars without fault within an months of purchase is going is 96, (i.e. 100 - 4 = 96).

Therefore, Pr. (Buying a car that will not develop fault) $= \dfrac{\text{Number of cars without fault}}{\text{Total number of cars}}$

$$= \dfrac{96}{100}$$

$$= \dfrac{24}{25} \quad \text{(In its lowest term after equal division by 4)}$$

6. In Mr. Smith's extended family, the number of males is 16, while the number of females is 14. Find the probability that Mr. Smith has:
(a) a male child
(b) a female child

Solution

(a) Total number of family members = 16 + 14 = 30

Therefore, Pr. (a male child) $= \dfrac{\text{Family members who are males}}{\text{Total number of family members}}$

$$= \dfrac{16}{30}$$

$$= \dfrac{8}{15}$$

(b) Pr. (a female child) $= \dfrac{\text{Family members who are females}}{\text{Total number of family members}}$

$$= \dfrac{14}{30}$$

$$= \dfrac{7}{15}$$

7. A survey shows that 36% of all women take size 8 shoes. What is the probability that Khan's grandmother takes size 8 shoes?

Solution

Pr. (Khan's grandmother takes size 8 shoes) = $\dfrac{36}{100}$ (Note that the total percentage is always 100%)

$= \dfrac{9}{25}$ (In its lowest term)

8. In a secondary school, 46 out of every 50 students are at least 130cm tall. What is the probability that a student chosen at random from the school is less than 130cm tall?

Solution

Total number of students for the sample = 50

Number of students who are at least 130cm tall = 46

Number of students who are less than 130cm tall = 50 - 46 = 4

Therefore, Pr. (a student less than 130cm tall) = $\dfrac{\text{Number of students less than 130cm tall}}{\text{Total number of students in the sample}}$

$= \dfrac{4}{50}$

$= \dfrac{2}{25}$

9. A number is chosen at random between 1 and 16, both inclusive. What is the probability that it is:

(a) even

(b) prime

(c) odd or prime

(d) divisible by 4

(e) a perfect square or a perfect cube

Solution

(a) Total numbers in all from 1 to 16 = 16

The even numbers are 2, 4, 6, 8, 10, 12, 14, 16

Therefore the number of even numbers is 8

Hence Pr. (even number selected) = $\dfrac{\text{Number of even numbers}}{\text{Total numbers in all}}$

$= \dfrac{8}{16}$

$= \dfrac{1}{2}$

(b) The prime numbers are 2, 3, 5, 7, 11, 13

Therefore the number of prime numbers is 6

Hence Pr. (prime number selected) = $\dfrac{\text{Number of prime numbers}}{\text{Total numbers in all}}$

$= \dfrac{6}{16}$

$= \dfrac{3}{8}$

(c) The odd numbers are 1, 3, 5, 7, 9, 11, 13, 15

The prime numbers are 1, 3, 5, 7, 11, 13

Since OR in probability means addition, then we add all the odd and prime numbers together, but we must not count any number twice. This gives 1, 3, 5, 7, 9, 11, 13, 15, which is a total of 8 numbers.

Hence Pr. (odd or prime number selected) = $\dfrac{\text{Number of odd and even numbers}}{\text{Total numbers in all}}$

$= \dfrac{8}{16}$

$= \dfrac{1}{8}$

(d) The numbers divisible by 4 are 4, 8, 12, 16

This gives a total of 4 numbers

Hence Pr. (a number divisible by 4) = $\dfrac{\text{The four numbers divisible by 4}}{\text{Total numbers in all}}$

$= \dfrac{4}{16}$

$= \dfrac{1}{4}$

(e) The perfect square numbers are 1, 4, 9, 16

The perfect cube numbers are 1, 8

Since OR in probability means addition, then we add all the set of values above without counting any number twice. This gives 1, 4, 8, 9, 15, which is a total of 5 numbers.

Hence Pr. (perfect square or perfect cube selected) = $\dfrac{15}{16}$

10. A letter is chosen at random from the alphabet. Find the probability that it is:

(a) T

(b) E or P

(c) not B or G

(d) either D, J, N, U, W or Y

(e) one of the letters of the word REJECTED

Solution

(a) There are 26 letters of the alphabet, out of which there is 1 T.

Therefore, Pr. (T) = $\dfrac{\text{Number of Ts}}{\text{Total numbers of alphabets}}$

$= \dfrac{1}{26}$

(b) Pr. (E or P) = $\dfrac{\text{Number of Es and Ps}}{\text{Total numbers of alphabets}}$

$= \dfrac{6}{26}$

$= \dfrac{1}{13}$

(c) Pr. (B or G) = $\dfrac{2}{26} = \dfrac{1}{13}$

Therefore, Pr. (not B or G) = 1 - Pr. (B or G)

$= 1 - \left(\dfrac{6}{13}\right)$

$= \dfrac{12}{13}$

(d) The letters D, J, N, U, W and Y makes a total of 6 letters.

Pr. (D, J, N, U, W or Y) = $\dfrac{6}{26}$

$= \dfrac{3}{13}$

(e) Writing the letters of the word REJECTED without repeating a letter gives REJCTD. This gives a total of 6 letters

Therefore Pr. (one of the letters of REJECTED) = $\dfrac{6}{26}$

$= \dfrac{3}{13}$

11. A letter is selected at random from the word PROBABILITY. What is the probability of selecting the letter B.

Solution

In this case the total letters of the word PROBABILITY gives 11. The repeated letters should be counted more than once since this is not a case of letter from the alphabet. In the 26 alphabet each letter appears once, that is why they are counted once. But in PROBABILITY (or other words that might be given) some letters appear more than once, hence they should be counted as many times as they appear.

In PROBABILITY, B appears 2 times.

Therefore, Pr. (selecting B) = $\dfrac{2}{11}$

Exercise 26

1. The table below give the number of students in each mark group in a class.

Mark	5	6	7	8	9	10
Number of students	3	6	2	4	1	4

If a student is chosen at random, find the probability that the student scored:

(a) 7 marks

(b) 6 marks or less

(c) at least 9 marks

(d) at most 8 marks

(e) 5 or 8 maks

2. The probability that a seed will germinate is $\frac{3}{4}$. What is the probability that it will not germinate?

3. A letter is chosen at random from the alphabet. Find the probability that it is one of the letters of the word: MATHEMATICS.

4. The probability that a man wins an election is $\frac{3}{5}$. What is the probability that he does not win.

5. Out of every 10 bulbs, 2 do not last long. What is the probability that a bulb will last long when lit?

6. In family, the number of males is 3, while the number of females is 2. Find the probability that another child born into the family is:

(a) a male child

(b) a female child

7. A survey shows that 44% of all women take size 7 shoes. What is the probability that a mother of two takes size 7 shoes?

8. In a secondary school, 30 out of every 100 students are at least 160cm tall. What is the probability that a student chosen at random from the school is less than 160cm tall?

9. A number is chosen at random between 1 and 20, both inclusive. What is the probability that it is:

(a) prime

(b) odd

(c) even or prime

(d) divisible by 3

(e) a number less than 10 or a perfect cube

10. A letter is chosen at random from the alphabet. Find the probability that it is:

(a) F

(b) M or Q or Y

(c) in the word COME

(d) either in the word BUT or in REMOVE

(e) one of the letters of the word SURPRISED

11. A letter is selected at random from the word RESPIRATION. What is the probability of selecting the letter I.

CHAPTER 27
PROBABILITY ON PACK OF PLAYING CARDS

A pack of playing cards contains 52 cards of 4 types. There are 13 clubs, 13 diamonds, 13 hearts and 13 spades. Each of the set of 13 cards contains Ace (A), 2, 3, 4, 5, 6, 7, 8, 9, 10, Jack (J), Queen (Q), and King (K). This means that out of the 52 cards, each card is four in number, i.e. Aces are 4 in number, 1s are 4 in number, 2s are 4 in number, 3s are 4 in number, 4s are 4 in number, 5s are 4 in number, 6s are 4 in number, 7s are 4 in number, 8s are 4 in number, 9s are 4 in number, 10s are 4 in number, Jacks are 4 in number, Queens are 4 in number, and Kings are 4 in number. Clubs and spades are black, diamonds and hearts are red. This means that there are 26 black cards and 26 red cards. This also means that out of the 4 Aces cards, 2 are black and 2 are red. Out of the four cards that are 1, two are black and two are red, out of the four cards that are 2, two are black and two are red, and so on.

Examples

1. A card is picked at random from a pack of playing cards. Find the probability of picking a spade. <u>Solution</u>

There are 13 spades in a pack of playing cards.

Therefore, Pr. (picking a spade) $= \dfrac{\text{Number of Spades}}{\text{Total numbers of cards}}$

$$= \dfrac{13}{52}$$

$$= \dfrac{1}{4} \quad \text{(In its lowest term)}$$

2. A card is picked at random from a pack of playing cards. Find the probability of picking a red card.

<u>Solution</u>

There are 26 red cards in a pack of playing cards.

Therefore, Pr. (picking a red card) $= \dfrac{\text{Number of red cards}}{\text{Total numbers of cards}}$

$$= \dfrac{26}{52}$$

$$= \dfrac{1}{2} \quad \text{(In its lowest term)}$$

3. A card is picked at random from a pack of playing cards. Find the probability of picking a red 5.

Solution

There are two red 5 cards in a pack of playing cards.

Therefore, Pr. (picking a red 5) = $\dfrac{\text{Number of red 5}}{\text{Total numbers of cards}}$

$\qquad = \dfrac{2}{52}$

$\qquad = \dfrac{1}{26}$ (In its lowest term)

4. A card is picked at random from a pack of playing cards. Find the probability of picking a 3.

Solution

There are 4 cards that are 3 in a pack of playing cards.

Therefore, Pr. (picking a 3) = $\dfrac{\text{Number of cards that are 3}}{\text{Total numbers of cards}}$

$\qquad = \dfrac{4}{52}$

$\qquad = \dfrac{1}{13}$ (In its lowest term)

5. A card is picked at random from a pack of playing cards. Find the probability of picking a black Ace.

Solution

There are 2 cards that are black ace in a pack of playing cards.

Therefore, Pr. (picking a black ace) = $\dfrac{\text{Number of cards that are black ace}}{\text{Total numbers of cards}}$

$\qquad = \dfrac{2}{52}$

$\qquad = \dfrac{1}{26}$ (In its lowest term)

6. A card is picked at random from a pack of playing cards. Find the probability of picking a card that is not a Jack.

Solution

(a) There are 4 cards that are Jacks.

Therefore, Pr. (picking a Jack) = $\dfrac{\text{Number of jacks}}{\text{Total numbers of cards}}$

$$= \frac{4}{52}$$

$$= \frac{1}{13}$$

Hence, Pr. (picking a card that in not a Jack) = 1 - Pr. (picking a Jack)

$$= 1 - \frac{1}{13}$$

$$= \frac{12}{13}$$

7. A card is picked at random from a pack of playing cards. Find the probability of picking

(a) a black or red card

(b) a 2 or a 5

(c) either a heart or the king of spades

(d) a club or a red Queen

(e) a diamond or a 9

(f) a 6 or a black card

Solution

(a) There are 26 black cards and 26 red card

Since or in probability means plus, then we have to add the numbers. This gives a total of: 26 + 26 = 52

Therefore, Pr. (picking a black or red card) = $\dfrac{\text{Number of black and red cards}}{\text{Total numbers of cards}}$

$$= \frac{52}{52}$$

$$= 1$$

(b) There are 4 cards that are 2, and 4 cards that are 5. This gives a total of 8 cards.

Therefore, Pr. (picking a 2 or a 5) = $\dfrac{8}{52}$

$$= \frac{2}{13}$$

(c) There are 13 cards that are Hearts, and 1 king that is a spade. This gives a total of 14 cards.

Therefore, Pr. (picking either a heart or the king of spades) = $\dfrac{14}{52}$

$$= \frac{7}{26}$$

(d) There are 13 cards that are club, and 2 cards that are red Queen, (i.e. the Queen of hearts and the queen of diamond). This gives a total of 15 cards.

Therefore, Pr. (picking a club or a red Queen) = $\dfrac{15}{52}$

(e) There are 13 cards that are diamonds, and 4 cards that are 9. But one of the 9 is in diamond and has already been counted among the 13 diamonds. So it must not be counted twice. Hence we count the other three 9 (each from clubs, hearts and spades). This will give a total of 16 (13 + 3) cards.

Therefore, Pr. (picking a diamond or a 9) = $\dfrac{16}{52}$

$= \dfrac{4}{13}$

(f) There are 4 cards that are 6, and 26 cards that are black. But two of the 26 black cards are among the four cards that are 6, and these two black 6 cards have already been counted among the 26 black cards. So they must not be counted twice. Hence we count the other two 6 cards that are red. This will give a total of 28 (26 + 2) cards.

Therefore, Pr. (picking a 6 or a back card) = $\dfrac{28}{52}$

$= \dfrac{7}{13}$

8. A card is picked at random from a pack of playing cards and then replaced. A second card is picked. What is the probability of picking:
(a) a 3 and a 10
(b) a queen and an ace
(c) two kings
(d) two red cards
(e) two cards of different colours
(f) two cards of the same colour

Solution

In probability problems, when two items are selected, it is important to logically analyse the situation when solving the problem. This will help you to know if addition (use of OR) is involved or multiplication (use of AND) is involved. For example, for a queen and a king to be selected, it simply means that, either the queen is selected first and then the king, or the king is selected first and then the queen. When this logical analysis is understood, then most questions in probability become easy to solve.

(a) There are four cards that are 3, and four cards that are 10

Therefore, Pr. (picking a 3) = $\dfrac{4}{52}$

$= \dfrac{1}{13}$

Similarly, Pr. (picking a 10) = $\dfrac{4}{52}$

$= \dfrac{1}{13}$

Recall that "and" in probability means multiplication.

The probability of picking a 3 and a 10 means that:

Either the first is a 3 AND the second is a 10, OR the first is a 10 AND the second is a 3.

This can be calculated by putting x in place of AND and + in place of OR in the above statement as follows:

Pr. (picking a 3) x Pr. (picking a 10) + Pr. (picking a 10) x Pr. (picking a 3)

$= (\dfrac{1}{13} \times \dfrac{1}{13}) + (\dfrac{1}{13} \times \dfrac{1}{13})$

$= \dfrac{1}{169} + \dfrac{1}{169}$

$= \dfrac{2}{169}$

Therefore, Pr. (picking a 3 and a 10) $= \dfrac{2}{169}$

(b) There are 4 cards that are queen, and 4 cards that are ace

Therefore, Pr. (picking a queen) $= \dfrac{4}{52}$

$= \dfrac{1}{13}$

Similarly, Pr. (picking an ace) $= \dfrac{4}{52}$

$= \dfrac{1}{13}$

The probability of picking a queen and an ace means that:

Either you first pick a queen AND then an ace, OR you first pick an ace AND then a queen.

This can be calculated by putting x in place of AND and + in place of OR in the above statement as follows:

Pr. (picking a queen) x Pr. (picking an ace) + Pr. (picking an ace) x Pr. (picking a queen)

$= (\dfrac{1}{13} \times \dfrac{1}{13}) + (\dfrac{1}{13} \times \dfrac{1}{13})$

$= \dfrac{1}{169} + \dfrac{1}{169}$

$= \dfrac{1}{169}$

Therefore, Pr. (picking a queen and an ace) $= \dfrac{2}{169}$

(c) There are four cards that are King

Therefore, Pr. (picking a king) $= \dfrac{4}{52}$

$= \dfrac{1}{13}$

The probability of picking two kings means that:

The first is a king AND the second is a king

= Pr. (picking a king) x Pr. (picking a king)

$= \dfrac{1}{13} \times \dfrac{1}{13}$

$= \dfrac{1}{169}$

Therefore, Pr. (picking two kings) = $\dfrac{1}{169}$

(d) There are 26 cards that are red

Therefore, Pr. (picking a red card) = $\dfrac{26}{52}$

$= \dfrac{1}{2}$

The probability of picking two red cards means that:

The first is a red card AND the second is a red card

= Pr. (picking a red card) x Pr. (picking a red card)

$= \dfrac{1}{2} \times \dfrac{1}{2}$

$= \dfrac{1}{4}$

Therefore, Pr. (picking two red cards) = $\dfrac{1}{4}$

(e) There are two colours of cards, red and black.

Therefore, Pr. (picking a red card) = $\dfrac{1}{2}$ (i.e from $\dfrac{26}{52}$ since there are 26 red cards)

Similarly, Pr. (picking a black card) = $\dfrac{1}{2}$ (i.e from $\dfrac{26}{52}$ since there are also 26 black cards)

The probability of picking two cards of different colours means that:

Either the first is a black card AND the second is a red card, OR the first is a red card AND the second is a black card.

This can be calculated by putting x in place of AND and + in place of OR in the above statement as follows:

Pr. (picking a black card) x Pr. (picking a red card) + Pr. (picking a red card) x Pr. (picking a black card)

$= (\dfrac{1}{2} \times \dfrac{1}{2}) + (\dfrac{1}{2} \times \dfrac{1}{2})$

$= \dfrac{1}{4} + \dfrac{1}{4}$

$= \dfrac{2}{4}$

$= \dfrac{1}{2}$

Therefore, Pr. (picking two cards of different colours) = $\dfrac{1}{2}$

(f) Pr. (picking two cards of the same colours) = 1 - Pr. (picking two cards of different colours)

$$= 1 - \frac{1}{2}$$

$$= \frac{1}{2}$$

Note that this can also be solved by using the logical process which is:

Either the first is red AND the second is red OR the first is black AND the second is black. This will also give $\frac{1}{2}$

9. Two cards are picked at random one after the other without replacement from a pack of playing cards. What is the probability of picking:

(a) a 5 and a 7

(b) a king and a jack

(c) two aces

(d) two diamond cards

(e) two black cards

(f) a red and a black card

(g) two cards of the same colours

Solution

This problem involves picking a card without replacement. This means that when one card is picked out, the total number of cards remaining in the pack become reduced to 51. That number of that particular type of card also reduces by 1.

(a) There are four cards that are 5. There are also four cards that are 7.

Hence the probability of picking a 5 and a 7 means that:

Either first picking a 5 AND then a 7, OR first picking a 7 AND then a 5.

Now, let us calculate each of the probabilities as follows:

Pr. (first card is a 5) = $\frac{4}{52}$ (There are four cards that are 5)

$= \frac{1}{13}$ (In its lowest term)

We now have 51 cards left in the pack.

Therefore, Pr. (second card is a 7) = $\frac{4}{51}$ (There are four cards that are 7, and a total of 51 cards remaining in the pack)

Or,

Pr. (first card is a 7) = $\frac{4}{52}$ (There are four cards that are 7)

$= \frac{1}{13}$ (In its lowest term)

We now have 51 cards left in the pack.

Therefore, Pr. (second card is a 5) = $\frac{4}{51}$ (There are four cards that are 5, and a total of 51 cards remaining in the pack)

Hence the probability of picking a 5 and a 7 means that:

Either first picking a 5 AND then a 7, OR first picking a 7 AND then a 5. Which is computed as:

Pr. (picking a 5 and a 7) = Pr. (first card is a 5) x Pr. (second card is a 7) + Pr. (first card is a 7) x Pr. (second card is a 5)

$$= (\frac{1}{13} \times \frac{4}{51}) + (\frac{1}{13} \times \frac{4}{51})$$

$$= \frac{4}{663} + \frac{4}{663}$$

$$= \frac{8}{663}$$

(b) There are four cards that are kings. There are also four cards that are jacks.

Now, let us calculate each of the probabilities as follows:

Pr. (first card is a king) = $\frac{4}{52}$ (There are four cards that are kings)

$\qquad = \frac{1}{13}$ (In its lowest term)

We now have 51 cards left in the pack.

Therefore, Pr. (second card is a jack) = $\frac{4}{51}$ (There are four cards that are jack, and a total of 51 cards remaining in the pack)

Or,

Pr. (first card is a jack) = $\frac{4}{52}$ (There are four cards that are jack)

$\qquad = \frac{1}{13}$ (In its lowest term)

We now have 51 cards left in the pack.

Therefore, Pr. (second card is a king) = $\frac{4}{51}$ (There are four cards that are king, and a total of 51 cards remaining in the pack)

Hence the probability of picking a king and a jack means that:

Either first picking a king AND then a jack, OR first picking a jack AND then a king. This is computed as:

Pr. (picking a king and a queen) = Pr. (first card is a king) x Pr. (second card is a jack) + Pr. (first card is a jack) x Pr. (second card is a king)

$$= (\frac{1}{13} \times \frac{4}{51}) + (\frac{1}{13} \times \frac{4}{51})$$

$$= \frac{4}{663} + \frac{4}{663}$$

$$= \frac{8}{663}$$

(c) There are 4 cards that are aces.

Hence the probability of picking two aces means that:

The first is an ace, and the second is an ace.

Now, let us calculate each of the probabilities as follows:

Pr. (first card is an ace) $= \frac{4}{52}$ (There are 4 cards that are aces)

$\qquad = \frac{1}{13}$ (In its lowest term)

We now have 3 aces left in the pack, and a total of 51 cards left in the pack.

Therefore, Pr. (second card is an ace) $= \frac{3}{51}$

Hence the probability of picking two aces is given by:

Pr. (picking two aces) = Pr. (first card is an ace) x Pr. (second card is an ace)

$\qquad = \frac{1}{13} \times \frac{3}{51}$

$\qquad = \frac{3}{663}$

(d) There are 13 cards that are diamonds.

Hence the probability of picking two diamonds means that:

The first is a diamond, and the second is a diamond.

Now, let us calculate each of the probabilities as follows:

Pr. (first card is a diamond) $= \frac{13}{52}$ (There are 13 cards that are diamonds)

$\qquad = \frac{1}{4}$ (In its lowest term)

We now have 12 diamonds left in the pack, and a total of 51 cards left in the pack.

Therefore, Pr. (second card is a diamond) $= \frac{1}{13}$

$\qquad = \frac{4}{17}$ (In its lowest term)

Hence the probability of picking two diamonds is given by:

Pr. (picking two diamonds) = Pr. (first card is a diamond) x Pr. (second card is a diamond)

$\qquad = \frac{1}{4} \times \frac{4}{17}$

$\qquad = \frac{4}{68}$

$\qquad = \frac{1}{17}$ (In its lowest term)

(e) There are 26 black cards.

Hence the probability of picking two black cards means that:

The first is a black card, and the second is a black card.

Now, let us calculate each of the probabilities as follows:

Pr. (first card is a black card) = $\dfrac{26}{52}$

$= \dfrac{1}{2}$ (In its lowest term)

We now have 25 black cards left in the pack, and a total of 51 cards left in the pack.

Therefore, Pr. (second card is a black card) = $\dfrac{25}{51}$

Hence the probability of picking two black cards is given by:

Pr. (picking two black cards) = Pr. (first card is a black card) x Pr. (second card is a black card)

$= \dfrac{1}{2} \times \dfrac{25}{51}$

$= \dfrac{25}{102}$

(f) The logical explanation for this situation is that:

Either the first card is red AND the second is black OR the first card is black and the second is red.

There are 26 red cards and also 26 black cards.

Now, let us calculate each of the probabilities as follows:

Pr. (first card is a red card) = $\dfrac{26}{52}$

$= \dfrac{1}{2}$ (In its lowest term)

We now have 51 cards left in the pack.

Therefore, Pr. (second card is a black card) = $\dfrac{26}{51}$ (There are 26 black cards, and a total of 51 cards remaining in the pack)

Or,

Pr. (first card is a black card) = $\dfrac{26}{52}$

$= \dfrac{1}{2}$ (In its lowest term)

We now have 51 cards left in the pack.

Therefore, Pr. (second card is a red card) = $\dfrac{26}{51}$ (There are 26 red cards, and a total of 51 cards remaining in the pack)

Hence the probability of picking a red card and a black card means that:

Either first picking a red card AND then a black card, OR first picking a black card AND then a red card. This is computed as:

Pr. (picking a red and black cards) = Pr. (first card is a red card) x Pr. (second card is a black card) + Pr. (first card is a black card) x Pr. (second card is a red card)

$$= (\frac{1}{2} \times \frac{26}{51}) + (\frac{1}{2} \times \frac{26}{51})$$

$$= \frac{26}{102} + \frac{26}{102}$$

$$= \frac{52}{102}$$

$$= \frac{26}{51}$$

(g) The logical explanation for this situation is that:

Either the first card is red AND the second is red OR the first card is black and the second is black.

There are 26 red cards and also 26 black cards.

Now, let us calculate each of the probabilities as follows:

Pr. (first card is a red card) $= \frac{26}{52}$

$$= \frac{1}{2} \text{ (In its lowest term)}$$

We now have 25 red cards left and a total of 51 cards left in the pack.

Therefore, Pr. (second card is a red card) $= \frac{25}{51}$

Or,

Pr. (first card is a black card) $= \frac{26}{102}$

$$= \frac{1}{2} \text{ (In its lowest term)}$$

We now have 25 black cards left and a total of 51 cards left in the pack.

Therefore, Pr. (second card is a black card) $= \frac{25}{51}$

Hence the probability of picking two cards of the same colour means that:

Either picking a red card AND then another red card, OR picking a black card AND then another black card. This is computed as:

Pr. (picking two cards of the same colour) = Pr. (first card is a red card) x Pr. (second card is a red card) + Pr. (first card is a black card) x Pr. (second card is a black card)

$$= (\frac{1}{2} \times \frac{25}{51}) + (\frac{1}{2} \times \frac{25}{51})$$

$$= \frac{25}{102} + \frac{25}{102}$$

$$= \frac{50}{102}$$

$$= \frac{25}{51}$$

Alternatively, this question can also be solved as follows:

Recall that question (f) above gives the probability of picking a red and a black card. This also means the probability of picking two cards of different colours.

Hence the probability of picking two cards of different colours as given in (f) above = $\dfrac{26}{51}$

Therefore, Pr. (picking two cards of the same colour) = 1 - Pr. (picking two cards of different colours)   (Note that they are opposite statements)

$$= 1 - \dfrac{26}{51}$$
$$= \dfrac{51-26}{51}$$
$$= \dfrac{25}{51} \quad \text{(As obtained before)}$$

10. If three cards are picked from a pack of playing cards with replacement, what is the probability if getting:
(a) at least two clubs
(b) at most two clubs

Solution
I am going to be using a special type of tree diagram without actually drawing the diagram.
Now, the total outcome in a selection of three items involving two events (i.e. a club or not a club) is given by:

2^n,

where n is the number of selection made.
In the question, n = 3, since three cards were picked.
Hence total outcome = 2^3 = 2 x 2 x 2

$\qquad$ = 8.

Now, in order to write out the outcomes, let us use the letter C to represent a club and letter N to represent not a club. Note that in tree diagrams like this, only two letters should be used in writing the outcomes since the question involves the picking of only one type of item (club). Hence the outcome is written as follows:

$\quad$ (CCC), (CCN), (CNC), (CNN), (NCC), (NCN), (NNC), (NNN)

Note that there are 8 ways of arranging the two letters in the brackets. There is no fast rule in carrying out the arrangement. You just have to make sure that no two brackets have the same arrangement of the letters. Also make sure the number of brackets is complete.

(a) In order to determine the probability of getting at least two clubs, we need to compute the probabilities of the brackets that contain at least 2 clubs. They are, (CCC), (CCN), (CNC), and (NCC). Note that at least two, means two and above, (i.e. two and three clubs in this case).
Hence the probability of getting at least two clubs = (CCC) or (CCN) or (CNC) or (NCC)
Now, let us compute each of the probabilities.

221

There are 13 clubs in a pack of cards, and there are 39 cards that are not club. Note that this is a case of with replacement, which means that the total number of cards in the pack is always complete. Hence:

(CCC) = Pr. (first card is a club) x Pr. (second card is a club) x Pr. (third card is a club)

$$= \frac{13}{52} \times \frac{13}{52} \times \frac{13}{52}$$

$$= \frac{1}{4} \times \frac{1}{4} \times \frac{1}{4}$$

$$= \frac{1}{64}$$

(CCN) = Pr. (first card is a club) x Pr. (second card is a club) x Pr. (third card is not a club)

$$= \frac{13}{52} \times \frac{13}{52} \times \frac{39}{52}$$ (Note that there are 39 cards that are not club)

$$= \frac{1}{4} \times \frac{1}{4} \times \frac{3}{4}$$

$$= \frac{3}{64}$$

(CNC) = Pr. (first card is a club) x Pr. (second card is not a club) x Pr. (third card is a club)

$$= \frac{13}{52} \times \frac{39}{52} \times \frac{13}{52}$$

$$= \frac{1}{4} \times \frac{3}{4} \times \frac{1}{4}$$

$$= \frac{3}{64}$$

(NCC) = Pr. (first card is not a club) x Pr. (second card is a club) x Pr. (third card is a club)

$$= \frac{39}{52} \times \frac{13}{52} \times \frac{13}{52}$$

$$= \frac{3}{4} \times \frac{1}{4} \times \frac{1}{4}$$

$$= \frac{3}{64}$$

Therefore, Pr. (getting at least two clubs) = (CCC) or (CCN) or (CNC) or (NCC)

$$= (CCC) + (CCN) + (CNC) + (NCC)$$

$$= \frac{1}{64} + \frac{3}{64} + \frac{3}{64} + \frac{3}{64}$$

$$= \frac{10}{64}$$

$$= \frac{5}{32}$$

(b) In order to determine the probability of getting at most two clubs, we need to compute the probabilities of the brackets that contain at most 2 clubs. From the outcome brackets given above, the ones that contain at most two clubs are, (CCN), (CNC), (CNN), (NCC), (NCN), (NNC), (NNN). Note that at most two, means two and below, (i.e. two, one and zero clubs in this case). Hence the probability of getting at most two clubs = (CCN) or (CNC) or (CNN) or (NCC) or (NCN) or (NNC) or (NNN)

Now, let us compute each of the probabilities. Hence:

(CCN) = $\dfrac{3}{64}$ (As calculated in (a) above)

(CNC) = $\dfrac{3}{64}$ (As calculated in (a) above)

(CNN) = Pr. (first card is a club) x Pr. (second card is not a club) x Pr. (third card is not a club)

$$= \dfrac{13}{52} \times \dfrac{39}{52} \times \dfrac{39}{52}$$
$$= \dfrac{1}{4} \times \dfrac{3}{4} \times \dfrac{3}{4})$$
$$= \dfrac{9}{64}$$

(NCC) = $\dfrac{3}{64}$ (As calculated in (a) above)

(NCN) = Pr. (first card is not a club) x Pr. (second card is a club) x Pr. (third card is not a club)

$$= \dfrac{39}{52} \times \dfrac{13}{52} \times \dfrac{39}{52}$$
$$= \dfrac{3}{4} \times \dfrac{1}{4} \times \dfrac{3}{4}$$
$$= \dfrac{9}{64}$$

(NNC) = Pr. (first card is not a club) x Pr. (second card is not a club) x Pr. (third card is a club)

$$= \dfrac{39}{52} \times \dfrac{39}{52} \times \dfrac{13}{52}$$
$$= \dfrac{3}{4} \times \dfrac{3}{4} \times \dfrac{1}{4}$$
$$= \dfrac{9}{64}$$

(NNN) = Pr. (first card is not a club) x Pr. (second card is not a club) x Pr. (third card is not a club)

$$= \dfrac{39}{52} \times \dfrac{39}{52} \times \dfrac{39}{52}$$
$$= \dfrac{3}{4} \times \dfrac{3}{4} \times \dfrac{3}{4}$$
$$= \dfrac{27}{64}$$

Therefore, Pr. (getting at most two clubs) = (CCN) or (CNC) or (CNN) or (NCC) or (NCN) or (NNC) or (NNN)

$$= (CCN) + (CNC) + (CNN) + (NCC) + (NCN) + (NNC) + (NNN)$$
$$= \dfrac{3}{64} + \dfrac{3}{64} + \dfrac{9}{64} + \dfrac{3}{64} + \dfrac{9}{64} + \dfrac{9}{64} + \dfrac{27}{64}$$
$$= \dfrac{63}{64}$$

Alternatively, a shorter method of solving this problem is as follows:

Pr. (getting at most two clubs) = 1 - Pr. (getting three clubs) (This is because the only item not in (b) is CCC. All 8 outcomes in brackets make a total probability of 1, but 7 of the outcomes are in (b), hence 1 minus the outcome not in (b) gives (b). This can also be expressed in fraction as, (b) = $\frac{7}{8}$ = $\frac{8}{8}$ - $\frac{1}{8}$). Therefore:

Pr. (getting at most two clubs) = 1 - Pr. (getting three clubs)

= 1 – CCC

= 1 - $\frac{1}{64}$ (Note that CCC computed in (a) = $\frac{1}{64}$)

= $\frac{63}{64}$ (As obtained before)

11. If three cards are chosen from a pack of playing cards without replacement, what is the probability of getting:
(a) at least two diamonds
(b) at most one diamond?

Solution

In order to write out the outcomes, let us use the letter D to represent a diamond and letter N to represent not a diamond.
Hence the outcomes are written as follows:

(DDD), (DDN), (DND), (DNN), (NDD), (NDN), (NND), (NNN)

(a) In order to determine the probability of getting at least two diamonds, we need to compute the probabilities of the brackets that contain at least 2 diamonds. They are, (DDD), (DDN), (DND), and (NDD).
Hence the probability of getting at least two diamonds = (DDD) or (DDN) or (DND) or (NDD)
Now, let us compute each of the probabilities.
There are 13 diamonds in a pack of cards, and there are 39 cards that are not diamonds. Note that this is a case of without replacement, which means that after each selection, both the total number of cards left and the number of the particular card picked, are reduced by 1. Hence:
(DDD) = Pr. (first card is a diamond) x Pr. (second card is a diamond) x Pr. (third card is a diamond)

= $\frac{13}{52}$ x $\frac{12}{51}$ x $\frac{11}{50}$ (Note that the number of diamond and the total number of card left, keep reducing by 1 after each selection)

= $\frac{1}{4}$ x $\frac{12}{51}$ x $\frac{11}{50}$

= $\frac{132}{10200}$

= $\frac{11}{850}$ (In its lowest term, after equal division by 12)

(DDN) = Pr. (first card is a diamond) x Pr. (second card is a diamond) x Pr. (third card is not a diamond)

$$= \frac{13}{52} \times \frac{12}{51} \times \frac{39}{50}$$ (Note that there are 39 cards that are not diamond)

$$= \frac{1}{4} \times \frac{4}{17} \times \frac{39}{50}$$

$$= \frac{156}{3400}$$

$$= \frac{39}{850}$$ (After equal division by 4)

(DND) = Pr. (first card is a diamond) x Pr. (second card is not a diamond) x Pr. (third card is a diamond)

$$= \frac{13}{52} \times \frac{39}{51} \times \frac{12}{50}$$

$$= \frac{1}{4} \times \frac{13}{17} \times \frac{6}{25}$$

$$= \frac{78}{1700}$$

$$= \frac{39}{850}$$

(NDD) = Pr. (first card is not a diamond) x Pr. (second card is a diamond) x Pr. (third card is a diamond)

$$= \frac{39}{52} \times \frac{13}{51} \times \frac{12}{50}$$

$$= \frac{3}{4} \times \frac{13}{51} \times \frac{6}{25}$$

$$= \frac{234}{5100}$$

$$= \frac{39}{850}$$

Therefore, Pr. (getting at least two diamonds) = (DDD) or (DDN) or (DND) or (NDD)

$$= (DDD) + (DDN) + (DND) + (NDD)$$

$$= \frac{11}{850} + \frac{39}{850} + \frac{39}{850} + \frac{39}{850}$$

$$= \frac{128}{850}$$

$$= \frac{64}{425}$$

(b) In order to determine the probability of getting at most one diamond, we need to compute the probabilities of the brackets that contain at most one diamond. From the outcome brackets given above, the ones that contain at most one diamond are, (DNN), (NDN), (NND), (NNN). Note that at most one, means one and below, (i.e. one and zero diamond in this case).

Hence the probability of getting at most one diamond = (DNN) + (NDN) + (NND) + (NNN)

Now, let us compute each of the probabilities. Hence:

225

(DNN) = Pr. (first card is a diamond) x Pr. (second card is not a diamond) x Pr. (third card is not a diamond)

$$= \frac{13}{52} \times \frac{39}{51} \times \frac{38}{50}$$

$$= \frac{1}{4} \times \frac{13}{17} \times \frac{19}{25}$$

$$= \frac{247}{1700}$$

(NDN) = Pr. (first card is not a diamond) x Pr. (second card is a diamond) x Pr. (third card is not a diamond)

$$= \frac{39}{52} \times \frac{13}{51} \times \frac{38}{50}$$

$$= \frac{3}{4} \times \frac{13}{51} \times \frac{19}{25}$$

$$= \frac{741}{5100}$$

$$= \frac{247}{1700} \quad \text{(In its lowest term after equal division by 3)}$$

(NND) = Pr. (first card is not a diamond) x Pr. (second card is not a diamond) x Pr. (third card is a diamond)

$$= \frac{39}{52} \times \frac{38}{51} \times \frac{13}{50}$$

$$= \frac{3}{4} \times \frac{38}{51} \times \frac{13}{50}$$

$$= \frac{1482}{10200}$$

$$= \frac{247}{1700} \quad \text{(After equal division by 6)}$$

(NNN) = Pr. (first card is not a diamond) x Pr. (second card is not a diamond) x Pr. (third card is not a diamond)

$$= \frac{39}{52} \times \frac{38}{51} \times \frac{37}{50}$$

$$= \frac{3}{4} \times \frac{38}{51} \times \frac{37}{50}$$

$$= \frac{4218}{10200}$$

$$= \frac{703}{1700} \quad \text{(After equal division by 6)}$$

Therefore, Pr. (getting at most one diamond) = (DNN) or (NDN) or (NND) or (NNN)

$$= \text{(DNN)} + \text{(NDN)} + \text{(NND)} + \text{(NNN)}$$

$$= \frac{247}{1700} + \frac{247}{1700} + \frac{247}{1700} + \frac{703}{1700}$$

$$= \frac{1444}{1700}$$

$$= \frac{361}{425}$$

Exercise 27

1. A card is picked at random from a pack of playing cards. Find the probability of picking a jack.

2. A card is picked at random from a pack of playing cards. Find the probability of picking a black 4.

3. A card is picked at random from a pack of playing cards. Find the probability of picking a red king.

4. A card is picked at random from a pack of playing cards. Find the probability of picking a either a black or red card.

5. A card is picked at random from a pack of playing cards. Find the probability of picking a black Queen.

6. A card is picked at random from a pack of playing cards. Find the probability of picking a card that is not an Ace.

7. A card is picked at random from a pack of playing cards. Find the probability of picking

(a) a queen or a king

(b) a 3 or a 9

(c) either a jack or the queen of diamonds

(d) a spade or a black 7

(e) a club or a red king

(f) a 2 or a red card

8. A card is picked at random from a pack of playing cards and then replaced. A second card is picked. What is the probability of picking:

(a) an 8 and a 5

(b) a black card and a 4

(c) two cards between 2 and 9 that have odd numbers

(d) two black cards

(e) two cards with the same number on them

(f) two cards with different number on them

9. Two cards are picked at random one after the other without replacement from a pack of playing cards. What is the probability of picking:

(a) a 4 and an ace

(b) a 2 and a 7

(c) two 8s

(d) two clubs

(e) two red cards

(f) a club and a diamond

(g) two cards that are queens

10. If three cards are picked from a pack of playing cards with replacement, what is the probability if getting:

227

(a) at least two 9s

(b) at most two 9s

11. If three cards are chosen from a pack of playing cards without replacement, what is the probability of getting:

(a) at least two kings

(b) at most one king?

CHAPTER 28
PROBABILITY ON TOSSING OF COINS

When a coin is tossed, the outcome can either be a head or a tail. However when two or more coins are tossed, the total outcome is obtained from 2^n, where n is the number of times the coin is tossed, or the number of coins tossed together.

Note that 'head' is the part of the coin that shows the person drawn on the coin, while the opposite side of the coin is called the 'tail'

Examples

1. A fair coin is tossed. What is the probability of getting:

(a) a head

(b) a tail

Solution

(a) There are only two possible outcomes. Head or tail.

Therefore, Pr. (getting a head) $= \dfrac{\text{Number of heads}}{\text{Total outcomes}}$

$= \dfrac{1}{2}$

(b) Pr. (getting a tail) $= \dfrac{\text{Number of tails}}{\text{Total outcomes}}$

$= \dfrac{1}{2}$

2. A coin is tossed two times. What is the probability of getting:

(a) a head and a tail

(b) at least a tail

(c) two heads

(d) two tails

(e) a head on the first toss, and a tail on the second toss.

Solution

The outcomes are written by using H for head and T for tail. The total number of outcomes will be $2^2 = 4$ (i.e. from 2^n, and n = 2 in this case)

The outcomes are: (HH), (HT), (TH), (TT).

(a) The outcomes with head and tail are (HT) and (TH). This gives 2 outcomes.

Therefore, Pr. (getting a head and tail) $= \dfrac{\text{Number of outcome with head and tail}}{\text{Total outcomes}}$

$$= \frac{2}{4}$$

$$= \frac{1}{2}$$

(b) The outcomes with at least a tail are (HT), (TH) and (TT). This gives 3 outcomes.

Therefore, Pr. (getting at least a tail) $= \dfrac{\text{Number of outcomes with at least a tail}}{\text{Total number of outcomes}}$

$$= \frac{3}{4}$$

(c) The outcome with two heads is (HH). This gives 1 outcome.

Therefore, Pr. (getting two heads) $= \dfrac{\text{Number of outcomes with heads}}{\text{Total number of outcomes}}$

$$= \frac{1}{4}$$

(d) The outcome with two tails is (TT). This gives 1 outcome.

Therefore, Pr. (getting two tails) $= \dfrac{\text{Number of outcomes with two tails}}{\text{Total number of outcomes}}$

$$= \frac{1}{4}$$

(e) The outcome with a head on the first toss, and a tail on the second toss is (HT). This gives 1 outcome

Therefore, Pr. (getting a head on the first toss, and a tail on the second toss) $= \dfrac{1}{4}$

3. A coin is tossed three times. What is the probability of getting:
(a) two heads and one tail
(b) at least one head
(c) three tails
(d) at least two heads
(e) a tail, a head and a tail

Solution
(a) The total number of outcomes will be $2^3 = 8$ (i.e. from 2^n, and n = 3 in this case)
The outcomes are: (HHH), (HTH), (HTT), (HHT), (THH), (THT), (TTH), (TTT). This gives a total of 8 outcomes

(a) The outcomes with two heads and one tail are (HTH), (HHT) and (THH). This gives 3 outcomes.
Therefore, Pr. (getting two heads and one tail) =

$$\frac{\text{Number of outcomes with two heads and one tail}}{\text{Total number of outcomes}}$$

$$= \frac{3}{8}$$

(b) The outcomes with at least one head are (HHH), (HTH), (HTT), (HHT), (THH), (THT) and (TTH). This gives 7 outcomes.

Therefore, Pr. (getting at least one head) = $\frac{\text{Number of outcomes with at least one head}}{\text{Total number of outcomes}}$

$$= \frac{7}{8}$$

(c) The outcome with three tails is (TTT). This gives 1 outcome.

Therefore, Pr. (getting three tails) = $\frac{1}{8}$

(d) The outcomes with at least two heads are (HHH), (HTH), (HHT) and (THH). This gives 4 outcomes.

Therefore, Pr. (getting at least two heads) = $\frac{\text{Number of outcomes with at least two heads}}{\text{Total number of outcomes}}$

$$= \frac{4}{8}$$
$$= \frac{1}{2}$$

(e) The outcome with a tail, a head and a tail is (THT). This is 1 outcome

Hence, Pr. (getting a tail, a head and a tail) = $\frac{1}{8}$

4. Four coins are tossed together. Find the probability of getting:
(a) two heads and two tails
(b) four tails
(c) at least three heads
(d) at least two heads
(e) one head

Solution
The total number of outcomes will be 2^4 = 16 (i.e. from 2^n, and n = 4 in this case)
The outcomes are: (HHHH), (HHHT), (HHTT), (HTTT), (THHH), (TTHH), (TTTH), (THTH), (HTHT), (HHTH), (THHT), HTTH), (TTHT), (THTT), (HTHH), (TTTT). This gives a total of 16 outcomes.

(a) The outcomes with two heads and two tails are (HHTT), (TTHH), (THTH), (HTHT), (THHT), (HTTH). This gives 6 outcomes.

Therefore, Pr. (getting two heads and two tails) =

$$\frac{\text{Number of outcomes with two heads and two tails}}{\text{Total number of outcomes}}$$

$$= \frac{6}{16}$$

$$= \frac{3}{8}$$

(b) The outcome with four tails is (TTTT). This gives 1 outcomes.

Therefore, Pr. (getting four tails) = $\frac{1}{16}$

(c) The outcomes with at least three heads are (HHHH), (HHHT), (THHH), (HHTH), (HTHH). This gives 5 outcomes.

Therefore, Pr. (getting at least three heads) = $\frac{\text{Number of outcomes with at least three heads}}{\text{Total number of outcomes}}$

$$= \frac{5}{16}$$

(d) The outcomes with at least two heads are (HHHH), (HHHT), (HHTT), (THHH), (TTHH), (THTH), (HTHT) (HHTH), (THHT), (HTTH), (HTHH). This gives 11 outcomes.

Therefore, Pr. (getting at least two heads) = $\frac{\text{Number of outcomes with at least two heads}}{\text{Total number of outcomes}}$

$$= \frac{11}{16}$$

(e) The outcomes with one head are, (HTTT), (TTTH), (TTHT), (THTT), . This gives 4 outcomes.

Therefore, Pr. (getting one head) = $\frac{\text{Number of outcomes with one head}}{\text{Total number of outcomes}}$

$$= \frac{4}{16}$$

$$= \frac{1}{4}$$

(5) A coin is tossed five times. Find the probability of getting at least one tail.

Solution

The total number of outcomes will be $2^5 = 32$

The only outcome without a tail is (HHHHH). This is an outcome of 1

Pr. (getting no tail, i.e. all head) = $\frac{1}{32}$

Therefore, Pr. (getting at least one tail) = 1 - Pr. (getting no tail)

$$= 1 - \frac{1}{32}$$

$$= \frac{31}{32}$$

Exercise 28

1. A fair coin is tossed. What is the probability of getting:

(a) a tail

(b) a head

(c) a tail or a head

2. A coin is tossed two times. What is the probability of getting:

(a) a tail and then a head

(b) at least a head

(c) two tails

(d) at least a tail

(e) a head on the first toss, and a tail on the second toss.

3. Three coins are tossed. What is the probability of getting:

(a) three heads

(b) at least one tail

(c) a head, a tail and then a head

(d) at least one head

(e) at least two heads

(f) at most two tails

4. Four coins are tossed together. Find the probability of getting:

(a) at least one head

(b) four heads

(c) at least two heads

(d) at most three tails

(e) two heads

(5) A coin is tossed five times. Find the probability of getting at least one head.

CHAPTER 29
PROBABILITY ON THROWING OF DICE

Examples

1. A fair die is thrown. Find the probability of getting:

(a) a 2

(b) a 5

(c) a 7

(d) a 4 or a 5

(e) a number less than 4

(f) an odd number

<u>Solution</u>

Note that a die has six faces numbered 1 to 6. That means that each number appears once.

(a) Pr. (getting a 2) = $\dfrac{\text{Number of faces having 2}}{\text{total number of faces}}$

$= \dfrac{1}{6}$

(b) Pr. (getting a 5) = $\dfrac{\text{Number of faces having 5}}{\text{total number of faces}}$

$= \dfrac{1}{6}$

(c) Pr. (getting a 7) = $\dfrac{\text{Number of faces having 7}}{\text{total number of faces}}$

$= \dfrac{0}{6}$

$= 0$ (This is a case of an impossible event)

(d) Pr. (getting a 4 or a 5) = $\dfrac{\text{Number of faces having 4 and having 5}}{\text{total number of faces}}$

$= \dfrac{2}{6}$

$= \dfrac{1}{3}$

(e) Pr. (getting a number less than 4) = $\dfrac{\text{Number of faces having numbers less than 4}}{\text{total number of faces}}$

$= \dfrac{3}{6}$ (Note that the faces with numbers less than 4 are 3, 2, and 1. This makes a total of 3 faces)

$= \dfrac{1}{2}$

(f) Pr. (getting an odd number) = $\dfrac{\text{Number of faces having odd numbers}}{\text{total number of faces}}$

$= \dfrac{3}{6}$ (Faces with odd numbers are 1, 3 and 5, i.e. three faces)

$= \dfrac{1}{2}$

2. A fair die is rolled once. What is the probability of getting:

(a) a number divisible by 3

(b) a multiple of 2

(c) at least 5

(d) at most 2

(e) a prime number or an even number

(f) either a number greater that 2 or a multiple of 4

Solution

(a) Pr. (getting a number divisible by 3) = $\dfrac{\text{Number of faces having numbers divisible by 3}}{\text{total number of faces}}$

$= \dfrac{2}{6}$ (Faces with numbers divisible by 3 are 3 and 6, i.e. 2 faces)

$= \dfrac{1}{3}$

(b) Pr. (getting a multiple of 2) = $\dfrac{\text{Number of faces having numbers that are multiple of 2}}{\text{total number of faces}}$

$= \dfrac{3}{6}$ (Faces with numbers that are multiple of 2 are 2, 4 and 6, i.e. 3 faces)

$= \dfrac{1}{2}$

(c) Pr. (getting at least 5) = $\dfrac{\text{Number of faces having numbers that are at least 5}}{\text{Total number of faces}}$

$= \dfrac{2}{6}$ (Faces with numbers that are at least 5 are 5 and 6, i.e. 2 faces)

$= \dfrac{1}{3}$

(d) Pr. (getting at most 2) = $\dfrac{\text{Number of faces having numbers that are at least 2}}{\text{Total number of faces}}$

$= \dfrac{2}{6}$ (Faces with numbers that are at most 2 are 1 and 2, i.e. 2 faces)

$= \dfrac{1}{3}$

(e) Pr. (getting a prime number or an even number) =

$$\frac{\text{Number of faces having prime numbers and even numbers}}{\text{Total number of faces}}$$

$= \frac{5}{6}$ (Faces with prime numbers are 2, 3, and 5. Faces with even numbers are 2, 4, 6. This will give a total of 5 faces because 2 which is both a prime and even number should be counted once)

Therefore, Pr. (getting a prime number or an even number) $= \frac{5}{6}$

(f) Pr. (getting either a number greater that 2 or a multiple of 4) =

$$\frac{\text{Number of faces having numbers greater than 2 and numbers that are multiple of 4}}{\text{Total number of faces}} = \frac{4}{6}$$ (Faces with numbers greater than 2 are 3, 4, 5 and 6. Faces with multiple of 4 is 4. This will give a total of 4 faces because 4 which appears in both events should be counted once)

Therefore, Pr. (getting either a number greater that 2 or a multiple of 4) $= \frac{4}{6}$

$= \frac{2}{3}$

3. A die is thrown and a coin is tossed. What is the probability of getting:
(a) a 3 and a head
(b) a tail and a prime number

Solution

(a) From the die, Pr. (getting a 3) $= \frac{1}{6}$

From the coin, Pr. (getting a head) $= \frac{1}{2}$

Since AND means multiplication in probability:

Therefore, Pr. (getting a 3 and a head) = Pr. (getting a 3) x Pr. (getting a head)

$= \frac{1}{6} \times \frac{1}{2}$

$= \frac{1}{12}$

(b) (a) From the coin, Pr. (getting a tail) $= \frac{1}{2}$

From the die, Pr. (getting a prime number) $= \frac{3}{6}$ (The prime numbers are 3, i.e. 2, 3 and 5)

$= \frac{1}{2}$

Therefore, Pr. (getting a tail and a prime number) = Pr. (getting a tail) x Pr. (getting a prime number)

$$= \frac{1}{2} \times \frac{1}{2}$$

$$= \frac{1}{4}$$

4. Two fair dice are thrown at the same time. Find the probability of getting:

(a) at least one six

(b) a sum of at least 10

(c) a sum of at most 5

(d) a sum less than 3

(e) a total of seven

(f) a sum that is either a prime number or a multiple of 3

(g) a sum that is either divisible by 3 or a multiple of 2

Solution

The outcome table is as shown below. The numbers in the bracket give the outcome from the first and second die respectively. Adding the numbers in the bracket will give the respective sum that will be obtained.

Number on second die

	+	1	2	3	4	5	6
	1	(1,1)	(1,2)	(1,3)	(1,4)	(1,5)	(1,6)
Number on	2	(2,1)	(2,2)	(2,3)	(2,4)	(2,5)	(2,6)
first die	3	(3,1)	(3,2)	(3,3)	(3,4)	(3,5)	(3,6)
	4	(4,1)	(4,2)	(4,3)	(4,4)	(4,5)	(4,6)
	5	(5,1)	(5,2)	(5,3)	(5,4)	(5,5)	(5,6)
	6	(6,1)	(6,2)	(6,3)	(6,4)	(6,5)	(6,6)

The outcome table above can be presented in a more direct form by adding the values in the brackets above to obtain the sum. This is as shown below. In the table below, the numbers in the brackets represent the numbers on each die. The numbers that are not in bracket are the outcomes from the sum of numbers on first and second dice.

Number on second die

+	(1)	(2)	(3)	(4)	(5)	(6)
(1)	2	3	4	5	6	7
(2)	3	4	5	6	7	8
(3)	4	5	6	7	8	9
(4)	5	6	7	8	9	10
(5)	6	7	8	9	10	11
(6)	7	8	9	10	11	12

Number on first die

Note that any of the tables above can be used to answer the questions asked above.

(a) The outcomes that can be obtained from getting at least a six are (6,1), (6,2), (6,3), (6,4), (6,5), (6,6), (1,6), (2,6), (3,6), (4,6), (5,6). They are from the first table. They are the outcomes from the 6 on the first die, and 6 on the second die respectively. The number of brackets from this outcome is 11 (when the brackets are counted). Note that the total outcomes from any of the two outcome tables above is 36. This is easily obtained from the second table by counting the numbers that are not in bracket.

Therefore, Pr. (getting at least a six) = $\dfrac{\text{Number of outcomes obtained when at least a six shows}}{\text{Total number of outcomes on the table}}$

$= \dfrac{11}{36}$

(b) A sum of at least 10 as shown on the second table above are, 10, 10, 10, 11, 11, and 12. This gives a total of 6 outcomes.

Therefore, Pr. (getting a sum of at least 10) = $\dfrac{6}{36}$ (Note that 36 is the total outcome)

$= \dfrac{1}{6}$

(c) A sum of at most 5 as shown on the second table above are, 5, 5, 5, 5, 4, 4, 4, 3, 3, and 2. This gives a total of 10 outcomes.

Therefore, Pr. (getting a sum of at most 5) = $\dfrac{10}{36}$

$= \dfrac{5}{18}$

(d) A sum less than 3 as shown on the second table above 2 only. This gives a total of 1 outcome.

Therefore, Pr. (getting a sum less than 3) = $\dfrac{1}{36}$

(e) A total of 7 as shown on the second table above appears 6 times. This gives a total of 6 outcomes.

Therefore, Pr. (getting a total of 7) = $\dfrac{6}{36}$

$\qquad = \dfrac{1}{6}$

(f) Sums which are prime numbers are 2, 3, 5, 7 and 11. Sums which are multiple of 3 are 3, 6, 9 and 12. Hence we are to count the outcomes from 2, 3, 5, 6, 7, 9, 11, and 12 (3 should be counted once). Hence, from the table, 2 appears 1 time, 3 appears 2 times, 5 appears 4 times, 6 appears 5 times, 7 appears 6 times, 9 appears 4 times, 11 appears 2 times, 12 appears 1 time. This gives a total outcome of 1 time + 2 times + 4 times + 5 times + 6 times + 4 times + 2 times + 1 time = 25. This is easier done on the table by counting all 2, 3, 5, 6, 7, 9, 11 and 12. It will also give a total of 25 outcomes.

Therefore, Pr. (getting a sum that is either a prime number or a multiple of 3) = $\dfrac{25}{36}$

(g) Sums which are divisible by 3 are 3, 6, 9 and 12. Sums which are multiples of 2 are 2, 4, 6, 8, 10 and 12. Hence we are to count the outcomes from 2, 3, 4, 6, 8, 9, 10 and 12 (6 and 12 which appear in both events should be counted once each). Hence, we go to the second table above and count all 2, 3, 4, 6, 8, 9, 10 and 12. It will give a total of 24 outcomes.

Therefore, Pr. (getting a sum that is either divisible by 3 or a multiple of 2) = $\dfrac{24}{36}$

$\qquad = \dfrac{2}{3}$

5. An unbiased die with faces numbered 1 to 6 is rolled twice. Find the probability that the product of the numbers obtained is:
(a) odd
(b) even
(c) 12
(d) prime
(e) either odd or a multiple of 5

Solution

The outcome table is as shown below. The numbers in brackets are the numbers on the die.

Number on second die

	x	(1)	(2)	(3)	(4)	(5)	(6)
	(1)	1	2	3	4	5	6
Number on	(2)	2	4	6	8	10	12
first die	(3)	3	6	9	12	15	18
	(4)	4	8	12	16	20	24
	(5)	5	10	15	20	25	30
	(6)	6	12	18	24	30	36

(a) All the odd numbers from the outcome table above are 1, 3, 3, 5, 5, 9, 15, 15, 25. This gives a total of 9 outcomes.

Therefore, Pr. (product of numbers is odd) = $\dfrac{9}{36}$ (Note that the total outcomes is 36)

$$= \dfrac{1}{4}$$

(b) Pr. (product of numbers is even) = 1 - Pr. (product of numbers is odd)

$$= 1 - \dfrac{1}{4}$$

$$= \dfrac{3}{4}$$

This can also be obtained by counting all the outcomes that are even numbers in the table above. Total even numbers is 27.

Hence, Pr. (product of numbers is even) = $\dfrac{27}{36}$

$$= \dfrac{3}{4} \quad \text{(As obtained before)}$$

(c) Pr. (product of numbers is 12) = $\dfrac{4}{36}$ (12 appears 4 times in the table)

$$= \dfrac{1}{9}$$

(d) All the prime numbers are, 2, 2, 3, 3, 5, 5. This gives a total outcomes of 6.

Therefore, Pr. (product of numbers is prime) $= \dfrac{6}{36}$

$$= \dfrac{1}{6}$$

(e) All products that are odd numbers are, 1, 3, 3, 5, 5, 9, 15, 15, 25. All products that are multiples of 5 are, 5, 5, 10, 10, 15, 15, 20, 20, 25, 30, 30.

They will both give a total outcome of 15. Note that, 5, 5, 15, 15, 25 are counted once under odd number. They should not be counted under multiples of 5, as this will result to double counting. Hence with this total outcome of 15,

Pr. (product of numbers is either odd or a multiple of 5) $= \dfrac{15}{36}$

$$= \dfrac{5}{12}$$

6. Three dice are thrown together. What is the probability of getting a total score of 10?

Solution.

If a die is thrown once, the total outcome is given by $6^1 = 6$. If two dice are thrown, the total outcome is $6^2 = 36$. Similarly, if three dice are thrown, the total outcome will be $6^3 = 216$. Now, for us to draw a table with 216 outcomes will be very tedious. So, a direct way of solving this problem will be to select the outcomes from each die that will result in a total score of 10. These outcomes are:

(6, 3, 1), (6, 2, 2), (5, 4, 1), (5, 3, 2), (4, 4, 2), (4, 3, 3)

Each of the brackets above can give us 6 outcomes. For example, the first bracket above can give us the following 6 outcomes:

(6, 3, 1): which means - First die shows 6, second die shows 3, third die shows 1

(6, 1, 3): which means - First die shows 6, second die shows 1, third die shows 3

(1, 6, 3): which means - First die shows 1, second die shows 6, third die shows 3

(1, 3, 6): which means - First die shows 1, second die shows 3, third die shows 6

(3, 1, 6): which means - First die shows 3, second die shows 1, third die shows 6

(3, 6, 1): which means - First die shows 3, second die shows 6, third die shows 1

Similarly, each of the other brackets can give us 6 outcomes.

Let us write out our outcome brackets again. They are, (6, 3, 1), (6, 2, 2), (5, 4, 1), (5, 3, 2), (4, 4, 2), (4, 3, 3)

When each of these brackets give us 6 outcomes, then we will obtain a total of 36 (i.e. 6 x 6) outcomes. Recall that our overall outcome table will give us a total of 216 (i.e. 6^3) outcomes.

Therefore, Pr. (getting a total score of 10) = $\dfrac{36}{216}$

$= \dfrac{1}{6}$

Exercise 29

1. A fair die is thrown once. Find the probability of getting:

(a) a 5

(b) a 1

(c) a 9

(d) a 2 or 3 or 6

(e) a number less than 6

(f) a prime or an even number

2. A fair die is rolled once. What is the probability of getting:

(a) a number divisible by 2

(b) a multiple of 3

(c) at least 2

(d) at most 3

(e) a perfect square or an odd number

(f) either a number greater that 5 or a multiple of 3

3. A die is thrown and a coin is tossed. What is the probability of getting:

(a) a 5 and a head

(b) a tail and a perfect cube

4. Two fair dice are thrown at the same time. Find the probability of getting:

(a) at least one four

(b) a sum of at least 6

(c) a sum of at most 10

(d) a sum less than 8

(e) a total of 12

(f) a sum that is either a perfect square or a multiple of 5

(g) a sum that is either divisible by 6 or a multiple of 4

5. An unbiased die with faces numbered 1 to 6 is rolled twice. Find the probability that the product of the numbers obtained is:

(a) prime

(b) divisible by 6

(c) 9

(d) a factor of 10

(e) either perfect cube or a multiple of 8

6. Three dice are thrown together. What is the probability of getting a total score of 11?

CHAPTER 30
MISCELLANEOUS PROBLEMS ON PROBABILITY

Examples

1. A box contains two green balls, three yellow balls and four white balls. A ball is picked at random from the box. What is the probability that it is:

(a) green

(b) yellow

(c) white

(d) blue

(e) not white

(f) either yellow or green

<u>Solution</u>

Total number of balls in the box = 2 + 3 + 4 = 9

(a) Pr. (that it is green) = $\dfrac{\text{Number of green balls}}{\text{Total number of balls in the box}}$

$\quad = \dfrac{2}{9}$

(b) Pr. (that it is yellow) = $\dfrac{\text{Number of yellow balls}}{\text{Total number of balls in the box}}$

$\quad = \dfrac{3}{9}$

$\quad = \dfrac{1}{3}$

(c) Pr. (that it is white) = $\dfrac{\text{Number of white balls}}{\text{Total number of balls in the box}}$

$\quad = \dfrac{4}{9}$

(d) There is no blue ball in the box.

Therefore, Pr. (that it is blue) = 0

(e) Pr. (that it is not white) = 1 - Pr. (that it is white)

$\quad = 1 - \dfrac{4}{9}$

$\quad = \dfrac{5}{9}$

(f) Pr. (that it is either yellow or green) = $\dfrac{\text{Number of yellow and green balls}}{\text{Total number of balls in the box}}$

$= \dfrac{3+2}{9}$

$= \dfrac{5}{9}$

Or,

Pr. (that it is either yellow or green) = Pr. (that it is yellow) + Pr. (that it is green)   (Since OR means addition)

$= \dfrac{1}{3} + \dfrac{2}{9}$

$= \dfrac{3+2}{9}$

$= \dfrac{5}{9}$ (As obtained before)

2. A letter is chosen at random from the word COMPUTER. What is the probability that it is:

(a) either in the word MORE or in the word CUT

(b) either in the word COPE or in the word CUTE

(c) neither in the word ROT nor in the word CUP

Solution

(a) The total number of letters in COMPUTER is 8 letters.

In the word MORE, the number of letters is 4, while in the word CUT, the number of letters is 3. They both give a total of 7 letters.

Therefore, Pr. (that it is either in the word MORE or in the word CUT) = $\dfrac{7}{8}$

(b) In the word COPE, the number of letters is 4, while in the word CUTE, the number of letters is 4. Without counting any letter twice (i.e. C and E), the two words give a total of 6 letters (i.e. C, O, P, E, U, T).

Therefore, Pr. (that it is either in the word COPE or in the word CUTE) = $\dfrac{6}{8}$ (The total number of letters in COMPUTER is 8 letters).

$= \dfrac{3}{4}$

(c) Out of the 8 letters in COMPUTER, the letters that are neither in the word ROT nor in the word CUP are letters M and E. They are 2 letters.

Therefore, Pr. (that it is neither in the word ROT nor in the word CUP) = $\dfrac{2}{8}$

$= \dfrac{1}{4}$

(3) In a college 80% of the boys and 45% of the girls can drive a car. If a boy and a girl are chosen at random, what is the probability that:

(a) both of then can drive a car |

(b) the boy cannot drive a car and the girl can drive a car

(c) neither of them can drive a car?

(d) one of them can drive a car

Solution

The probabilities are given in percentage. Hence the total for each probability is 100%

Therefore, Pr. (a boy can drive a car) = $\dfrac{80}{100}$

$= \dfrac{4}{5}$

Pr. (a boy cannot drive a car) = $\dfrac{20}{100}$ (i.e. 100 - 80 = 20)

$= \dfrac{1}{5}$ (Can also be obtained from $1 - \dfrac{4}{5}$)

Similarly, Pr. (a girl can drive a car) = $\dfrac{45}{100}$

$= \dfrac{9}{20}$ (After equal division by 5)

Pr. (a girl cannot drive a car) = $1 - \dfrac{9}{20}$)

$= \dfrac{11}{20}$

(a) Therefore, Pr. (both of them can drive a car) = Pr. (a boy can drive a car) AND Pr. (a girl can drive a car)

$=$ Pr. (a boy can drive a car) x Pr. (a girl can drive a car)

$= \dfrac{4}{5} \times \dfrac{9}{20}$

$= \dfrac{36}{100}$

$= \dfrac{9}{25}$

(b) Pr. (the boy cannot drive a car and the girl can drive a car) = Pr. (a boy cannot drive a car) AND Pr. (a girl can drive a car)

$=$ Pr. (a boy cannot drive a car) x Pr. (a girl can drive a car)

$= \dfrac{1}{5} \times \dfrac{9}{20}$

$= \dfrac{9}{100}$

(c) Pr. (neither of them can drive a car) = Pr. (a boy cannot drive a car) AND Pr. (a girl cannot drive a car)

= Pr. (a boy cannot drive a car) x Pr. (a girl cannot drive a car)

$$= \frac{1}{5} \times \frac{11}{20}$$

$$= \frac{11}{100}$$

(d) Since we do not know which of then can drive a car, then this case is logically explained as follows:

Pr. (one of them can drive a car) = either the boy can drive a car AND the girl cannot drive a car OR the girl can drive a car AND the boy cannot drive a car.

This in now calculated as follows:

Pr. (one of them can drive a car) = Pr. (the boy can drive a car) x Pr. (the girl cannot drive a car) + Pr. (the girl can drive a car) x Pr. (the boy cannot drive a car)

$$= (\frac{4}{5} \times \frac{11}{20}) + (\frac{9}{20} \times \frac{1}{5})$$

$$= \frac{11}{25} + \frac{9}{100}$$

$$= \frac{44 + 9}{100}$$

$$= \frac{53}{100}$$

4. The probability of a seed germinating is $\frac{2}{5}$. If three of the seeds are planted, what is the probability that:

(a) none will germinate

(b) at least one will germinate

(c) at least one will not germinate

(d) only one will germinate

Solution

This is a case of selection of three items from two possible events. We are going to write our outcomes in bracket like a tree diagram method. In order to write out the outcomes, let us use the letter G to represent germinate and letter N to represent not germinate.

Hence the outcomes are written as follows:

(GGG), (GGN), (GNG), (GNN), (NGG), (NGN), (NNG), (NNN)

(a) The probability that none will germinate is given by (NNN).

From the question, the probability that a seed germinate, G = $\frac{2}{5}$. Therefore the probability that it will

not germinate, N = 1 - G = 1 - $\frac{2}{5}$ = $\frac{3}{5}$

Hence, $G = \dfrac{2}{5}$, $N = \dfrac{3}{5}$

Therefore, Pr. (that none will germinate) = (NNN)

$$= \dfrac{3}{5} \times \dfrac{3}{5} \times \dfrac{3}{5}$$

$$= \dfrac{27}{125}$$

(b) The outcomes of the probability that at least one will germinate are, (GGG), (GGN), (GNG), (GNN), (NGG), (NGN), (NNG). Hence we can compute each of the outcomes and add them together. But this will be tedious. An easier way of solving this problem is as explained below. The difference between the outcome in question (a) and (b) is (NNN). This shows that subtracting (NNN) from the total probability will give us the outcomes in question (b). Recall that the total of any probability is 1. Therefore, 1 - (NNN) = outcomes in (b)

Hence, Pr. (that at least one will germinate) = 1 - (NNN)

$$= 1 - \dfrac{27}{125} \quad \text{[Note that (NNN)} = \dfrac{27}{125} \text{ as calculated in question (a)]}$$

$$= \dfrac{108}{125}$$

(c) The outcomes of the probability that at least one will not germinate are, (GGN), (GNG), (GNN), (NGG), (NGN), (NNG), (NNN). Similar to (b) above, the difference between this outcomes of this question and the overall outcomes is (GGG).

Therefore, Pr. (that at least one will not germinate) = 1 - (GGG)

Let us calculate (GGG) as follows:

Pr. [that all three will germinate, i.e. (GGG)] $= \dfrac{2}{5} \times \dfrac{2}{5} \times \dfrac{2}{5}$

$$= \dfrac{8}{125}$$

Therefore, Pr. (that at least one will not germinate) = 1 - (GGG)

$$= 1 - \dfrac{8}{125}$$

$$= \dfrac{117}{125}$$

(d) The outcomes of the probability that only one will germinate are, (GNN), (NGN), (NNG). Hence we will calculate each of these outcomes and add them together.

(GNN) = Pr. (that the first will germinate) x Pr. (that the second will not germinate) x Pr. (that the third will not germinate)

$$= \dfrac{2}{5} \times \dfrac{3}{5} \times \dfrac{3}{5}$$

$$= \dfrac{18}{125}$$

(NGN) $= \dfrac{3}{5} \times \dfrac{2}{5} \times \dfrac{3}{5}$

$$= \frac{18}{125}$$

$(NNG) = \frac{3}{5} \times \frac{3}{5} \times \frac{2}{5}$

$$= \frac{18}{125}$$

Therefore, Pr. (that only one will germinate) $= \frac{18}{125} + \frac{18}{125} + \frac{18}{125}$

$$= \frac{54}{125}$$

5. When children are born, they are equally likely to be boys or girls. What is the probability that in a family of four children:

(a) three are boys and one is a girl

(b) at least two are girls

(c) two are boys and two are girls

(d) the first and second born are girls

Solution

Since children are equally likely to be boys or girls, it means that the probability of having a boy is $\frac{1}{2}$,

and the probability of having a girl is also $\frac{1}{2}$. This is similar to the case of tossing a coin (i.e. $\frac{1}{2}$ for head and $\frac{1}{2}$ for tail).

Therefore, the case of a family of four children is like when four coins are tossed. Refer to the example on tossing four coins in our previous chapter.

Let us use B for boy and G for girl to write out the total outcomes of 16 (i.e. $2^4 = 16$) as shown below.

The outcomes are: (BBBB), (BBBG), (BBGG), (BGGG), (GBBB), (GGBB), (GGGB), (GBGB), (BGBG), (BBGB), (GBBG), (BGGB), (GGBG), (GBGG), (BGBB), (GGGG). This gives a total of 16 outcomes.

(a) The outcomes that the children are three boys and one girl are, (BBBG), (GBBB), (BBGB), (BGBB). This gives 4 outcomes.

Therefore, Pr. (three are boys and one is a girl) $= \frac{4}{16}$

$$= \frac{1}{4}$$

(b) The outcomes that the children are at least two girls are, (BBGG), (BGGG), (GGBB), (GGGB), (GBGB), (BGBG), (GBBG), (BGGB), (GGBG), (GBGG), (GGGG). This gives 11 outcomes.

Therefore, Pr. (at least two are girls) = $\dfrac{11}{16}$

(c) The outcomes that the children are two boys and two girls are, (BBGG), (GGBB), (GBGB), (BGBG), (GBBG), (BGGB). This gives 6 outcomes.

Therefore, Pr. (two are boys and two are girls) = $\dfrac{6}{16}$

$\qquad = \dfrac{3}{8}$

(d) The outcomes that the first and second born are girls are, (GGBB), (GGGB), (GGBG), (GGGG). This gives 4 outcomes.

Therefore, Pr. (the first and second born are girls) = $\dfrac{4}{16}$

$\qquad = \dfrac{1}{4}$

6. A bag contains three blue balls, four red balls and five white balls. Three balls are removed from the bag without replacement. What is the probability of getting:
(a) a white, blue and red balls in that order
(a) one of each colour
(c) at least two white balls

Solution
The total number of balls in the bag = 3 + 4 + 5 = 12

(a) A white, blue and red balls in that order means that the first is white, the second is blue and the third is red. This can be represented as (WBR).
Note that this is a case of without replacement. Hence after each ball is removed, the total number of ball remaining and the number of the particular ball removed are both reduced by one.

Therefore, Pr. (getting a white, blue and red balls, i.e. WBR) = $\dfrac{5}{12} \times \dfrac{3}{11} \times \dfrac{4}{10}$. (Notice how the total balls is reduced by 1 after each ball is removed from the bag.

$\qquad = \dfrac{60}{1320}$

$\qquad = \dfrac{1}{22} \qquad$ (After equal division by 60)

(b) Let B represent blue, R represent red and W represent white. Then the outcomes for getting one of each colour are given by: (BRW), (BWR), (RBW), (RWB), (WBR), (WRB).

Let us now calculate each of them.

(BRW) = Pr. (First is blue) x Pr. (Second is red) x Pr. (Third is white)

$$= \frac{3}{12} \times \frac{4}{11} \times \frac{5}{10}$$

$$= \frac{1}{4} \times \frac{4}{11} \times \frac{1}{2}$$

$$= \frac{4}{88}$$

$$= \frac{1}{22}$$

Similarly, each of the other five outcomes, i.e. (BWR), (RBW), (RWB), (WBR), (WRB), will each give us a value of $\frac{1}{22}$ when calculated. This is because each is obtained by multiplying 3 x 4 x 5, to give the numerator, and 12 x 11 x 10, to give the denominator, which simplifies to $\frac{1}{22}$.

Therefore, Pr. (getting one of each colour) $= \frac{1}{22} + \frac{1}{22} + \frac{1}{22} + \frac{1}{22} + \frac{1}{22} + \frac{1}{22}$

$$= \frac{6}{22}$$

$$= \frac{3}{11}$$

(c) Let us write out a different outcome for this problem. Since we are concerned about one colour, we are going to use W to represent white colour, and N to represent not a white colour. This will give us 8 outcomes in brackets as usual. The outcomes are:

(WWW), (WWN), (WNW), (WNN), (NWW), (NWN), (NNW), (NNN).

The outcomes representing at least two white balls are: (WWW), (WWN), (WNW), (NWW).

Number of white balls is 5. Therefore number of balls that are not white = 12 - 5 = 7, or blue + red = 3 + 4 = 7. (Blue and red ball are the balls that are not white balls).

Let us now calculate each of the outcomes above as follows:

(WWW) = Pr. (first is white) x Pr. (second is white) x Pr. (third is white)

$$= \frac{5}{12} \times \frac{4}{11} \times \frac{3}{10}$$ (Take note of the reduction in the white balls and total number of balls as each ball is removed from the bag)

$$= \frac{60}{1320}$$

$$= \frac{1}{22}$$

(WWN) $= \frac{5}{12} \times \frac{4}{11} \times \frac{7}{10}$ (Note that there are 7 balls that are not white)

$$= \frac{140}{1320}$$

$$= \frac{7}{66}$$

(WNW) $= \frac{5}{12} \times \frac{7}{11} \times \frac{4}{10}$

$$= \frac{140}{1320}$$

$$= \frac{7}{66}$$

$$(NWW) = \frac{7}{12} \times \frac{5}{11} \times \frac{4}{10}$$

$$= \frac{140}{1320}$$

$$= \frac{7}{66}$$

Therefore, Pr. (getting at least two white balls) = (WWW) or (WWN) or (WNW) or (NWW)

$$= (WWW) + (WWN) + (WNW) + (NWW)$$

$$= \frac{1}{22} + \frac{7}{66} + \frac{7}{66} + \frac{7}{66}$$

$$= \frac{3 + 7 + 7 + 7}{66}$$

$$= \frac{24}{66}$$

$$= \frac{4}{11}$$

7. A committee consist of 6 men and 4 women. A subcommittee made up of three members is randomly chosen from the committee members. What is the probability that:

(a) they are all men

(b) two of them are women?

Solution

Let us write out the outcome for this problem. Let M represent man, and W represent woman. This will give us 8 outcomes in brackets as usual. The outcomes are:

(WWW), (WWM), (WMW), (WMM), (MWW), (MWM), (MMW), (MMM).

(a) The total members in the committee are: 6 + 4 = 10.

The outcomes representing all men is (MMM)

Therefore, Pr. (they are all men, i.e. MMM) = Pr. (first is a man) x Pr. (second is a man) x Pr. (third is a man)

$$= \frac{6}{10} \times \frac{5}{9} \times \frac{4}{8}$$ (Notice the reduction in the number of men and people left, as each

member is chosen from the committee).

$$= \frac{130}{720}$$

$$= \frac{13}{72}$$

(b) The outcomes showing that two of them are women are: (WWM), (WMW), (MWW)

Let us calculate each of them as follows:

(WWM) = Pr. (the first is a woman) x Pr. (the second is a woman) x Pr. (the third is a man)

$$= \frac{4}{10} \times \frac{3}{9} \times \frac{6}{8}$$

$$= \frac{72}{720}$$

$$= \frac{1}{10}$$

$$(WMW) = \frac{4}{10} \times \frac{6}{9} \times \frac{3}{8}$$

$$= \frac{72}{720}$$

$$= \frac{1}{10}$$

$$(MWW) = \frac{6}{10} \times \frac{4}{9} \times \frac{3}{8}$$

$$= \frac{72}{720}$$

$$= \frac{1}{10}$$

Therefore, Pr. (two of them are women) = (WWM) or (WMW) or (MWW)

$$= (WWM) + (WMW) + (MWW)$$

$$= \frac{1}{10} + \frac{1}{10} + \frac{1}{10}$$

$$= \frac{3}{10}$$

8. A box contains seven blue pens and three red pens. Three pens are picked one after the other without replacement. Find the probability of picking:

(a) two blue pens

(b) at least two red pens

(c) at most two blue pens

Solution

Let B represent blue pen, and R represent red pen. The outcomes are:

(BBB), (BBR), (BRB), (BRR), (RBB), (RBR), (RRB), (RRR).

The total number of pens = 7 + 3 = 10

(a) The outcomes showing two blue pens are: (BBR), (BRB), (RBB)

Let us calculate each of them as follows:

(BBR) = Pr. (the first is a blue pen) x Pr. (the second is a blue pen) x Pr. (the third is a red pen)

$$= \frac{7}{10} \times \frac{6}{9} \times \frac{3}{8}$$

$$= \frac{126}{720}$$

$$= \frac{7}{40} \quad \text{(In its lowest term after equal division by 18)}$$

$$\text{(BRB)} = \frac{7}{10} \times \frac{3}{9} \times \frac{6}{8}$$

$$= \frac{126}{720}$$

$$= \frac{7}{40}$$

Also, (RBB) = $\frac{7}{40}$ (Similar to the once above)

Therefore, Pr. (picking two blue pens) = $\frac{7}{40} \times \frac{7}{40} \times \frac{7}{40}$

$$= \frac{21}{40}$$

(b) The outcomes representing at least two red pens are: (RRR), (RRB), (RBR), (BRR)

Let us now calculate each of the outcomes as follows:

(RRR) = Pr. (first is a red pen) x Pr. (second is a red pen) x Pr. (third is a red pen)

$$= \frac{3}{10} \times \frac{2}{9} \times \frac{1}{8} \quad \text{(Take note of the reduction in the red pens and total number of pens as}$$

each pen is picked from the box)

$$= \frac{6}{720}$$

$$= \frac{1}{120}$$

$$\text{(RRB)} = \frac{3}{10} \times \frac{2}{9} \times \frac{7}{8}$$

$$= \frac{42}{720}$$

$$= \frac{7}{120}$$

Hence, (RBR) = $\frac{7}{120}$ (This is similar to the one above)

And, (BRR) = $\frac{7}{120}$ (Same reason as above)

Therefore, Pr. (picking at least two red pens) = $\frac{1}{120} + \frac{7}{120} + \frac{7}{120} + \frac{7}{120}$

$$= \frac{1+7+7+7}{120}$$

$$= \frac{22}{120}$$

$$= \frac{11}{60}$$

(c) The outcomes that represent picking at most two blue pens are: (BBR), (BRB), (BRR), (RBB), (RBR), (RRB), (RRR). Note that at most two blue pens means 2, 1 or 0 blue pens.

Notice that there is only (BBB) missing from this outcome. This shows that it can be obtained by: total probability - (BBB). Which is: 1 - (BBB).

Let us calculate (BBB) as follows:

(BBB) = Pr. (first is a blue pen) x Pr. (second is a blue pen) x Pr. (third is a blue pen)

$$= \frac{7}{10} \times \frac{6}{9} \times \frac{5}{8}$$

$$= \frac{210}{720}$$

$$= \frac{7}{24} \quad \text{(After equal division by 30)}$$

Therefore, Pr. (picking at most two blue pens) = 1 - (BBB)

$$= 1 - \frac{7}{24}$$

$$= \frac{17}{24}$$

Exercise 30

1. A box contains 5 green balls, 8 yellow balls and 7 white balls. A ball is picked at random from the box. What is the probability that it is:

(a) green

(b) yellow

(c) white

(d) blue

(e) not white

(f) either yellow or green

2. A letter is chosen at random from the word NORMADIC. What is the probability that it is:

(a) either in the word MAD or in the word CORN

(b) either in the word NORM or in the word DAM

(c) neither in the word RID nor in the word CAN

(3) In a college 20% of the boys and 8% of the girls who had graduated from the college, graduated with distinction since the inception of the college. If a boy and a girl are chosen at random, what is the probability that:

(a) both of them will graduate with distinction

(b) the boy will not and the girl will graduate with distinction

(c) neither of them will graduate with distinction?

(d) one of them will graduate with distinction

4. The probability of a seed germinating is $\frac{1}{4}$. If three of the seeds are planted, what is the probability that:
(a) none will germinate
(b) at least one will germinate
(c) at least one will not germinate
(d) only one will germinate

5. When parents who are carriers of sickle cell disorder get married, they are equally likely to give birth to normal child and sick child. What is the probability that in a family of three children:
(a) two are normal and one is sick
(b) at least two are sick
(c) one is normal and two are sick
(d) the first is sick
(e) at most one is normal

6. A box contains six blue balls, three red balls and five white balls. Three balls are removed from the bag without replacement. What is the probability of getting:
(a) a white, blue and red balls in that order
(a) one of each colour
(c) at least two white balls

7. A committee consist of 4 men and 2 women. A subcommittee made up of two members is randomly chosen from the committee members. What is the probability that:
(a) they are all men
(b) one of them is a woman?

8. A bag contains 5 blue balls and seven red balls. Three balls are picked one after the other without replacement. Find the probability of picking:
(a) two blue balls
(b) at least two red balls
(c) at most two blue balls

ANSWERS TO EXERCISES

Exercise 1

1. 9   2. $-\dfrac{8}{5}$ 3. 9   4. 14   5. -14 6. $\dfrac{2}{5}$ 7. $\dfrac{1}{3}$ 8. -1 9. -15

10. 10   11. $\dfrac{1}{4}$ 12. 27   13. 12   14. $\dfrac{1}{10}$ 15. $\dfrac{3}{7}$ 16. Continuous

17. Continuous   18. Discontinuous   19. Not continuous (Discontinuous)

20. Not continuous   21. Continuous   22. Not continuous   23. Not continuous

24(a) 8   (b) 5   26. $\dfrac{7}{2}$ 27. $\dfrac{1}{3}$ 28. 1   29. $-\dfrac{1}{2}$ 30. $2\dfrac{1}{4}$

Exercise 2

1. 2   2. $2x$ 3. $-3x^{-4}$ or $-\dfrac{3}{x^4}$ 4(a) $10x + 5h$ (b) $10x$

5(a) $27x^2 + 27xh + 9h^2$ (b) $27x^2$ 6(a) $\dfrac{2x^3 + 3x^2\Delta x + x(\Delta x)^2 + 2}{x(x + \Delta x)}$ (b) $2x + \dfrac{2}{x^2}$

7. $6x - 10$ 8. $1 - \dfrac{3}{x^2}$ 9. $5 - 6x$ 10. $2 + \dfrac{1}{5} = 2\dfrac{1}{5}$

Exercise 3

1(a) $40x^4$ (b) $2x^4$ (c) $\dfrac{1}{3x^{\frac{2}{3}}}$ or $\dfrac{1}{3\sqrt[3]{x^2}}$ (d) $\dfrac{1}{x^{\frac{6}{7}}}$ or $\dfrac{1}{\sqrt[7]{x^6}}$ (e) $-\dfrac{5}{8x^{\frac{13}{8}}}$ or $-\dfrac{5}{8\sqrt[8]{x^{13}}}$

(f) $-\dfrac{5}{x^{\frac{7}{2}}}$ or $-\dfrac{5}{\sqrt{x^7}}$ 2. (a) $10x^4 - 12x^3 - 12x^2 + 10x - 6$ (b) $7x^6 + 8x^3 + \dfrac{3}{x^2}$

(c) $18x^5 - 4x^3 - 5 - \dfrac{2}{x^2} + \dfrac{3}{x^4}$ (d) $\dfrac{5}{4\sqrt[4]{x^3}} - \dfrac{5}{3\sqrt[3]{2x^4}}$ 3. $8(2x-5)^3$ 4. $\dfrac{-6}{(x^3 - 7)^3}$

5. $\dfrac{6x^2 + 7}{2(2x^3 + 7x)^{\frac{1}{2}}}$ 6. $5(21x^2 - 2x)(7x^3 - x^2 + 3)^4$ 7. $\left(9 + \dfrac{2}{x^2}\right)\left(3x - \dfrac{2}{3x}\right)^2$

8. $1 + \dfrac{9(6x - 1)}{(3x^2 - x - 10)^2}$ 9. $\dfrac{-4x^3}{\sqrt{1 - 2x^4}}$ 10. $\dfrac{-5x^2}{\sqrt[3]{(5x^3 - 1)^4}}$

Exercise 4

1. $12x - 1$ 2. $12x^3 + 50x$ 3. $\dfrac{15x + 30}{2\sqrt{3 + x}}$ 4. $(7x^2 + 30 - 7)(x^2 - 7)^2$

5. $48x^3 + 24x^2 - 6x - 20$ 6. $\dfrac{\sqrt{2}\,[(3x^2 - 1)^3 + 36x^2(9x^4 - 6x^2 + 1)]}{2\sqrt{x}}$

7. $\dfrac{15 - 7x}{4(x+3)^{\frac{1}{4}}}$ or $\dfrac{15 - 7x}{4\sqrt[4]{x + 3}}$ 8. $12x^3 + 9x^2 - 46x - 5$ 9. $15x^4 - 12x^3 - 3x^2 + 6x - 2$

10. $\dfrac{12x + 1}{2}$ 11. $\dfrac{9x^{\frac{7}{2}}}{2}$ or $\dfrac{9\sqrt{x^7}}{2}$ 12. $\dfrac{4x^3(7x - 33)}{3\sqrt[3]{2x - 11}}$

257

13. $-30x^5 - 175x^4 + 20x^3 + 102x^2 - 14x + 1$ 14. $75x^4 + 28x^3 + 3x^2 + 28x - 7$

15. $-\dfrac{2}{x^2} + \dfrac{12}{x^5} - \dfrac{25}{x^6}$

Exercise 5

1. $\dfrac{x^2 - 2x + 4}{(x-1)^2}$ 2. $\dfrac{8x^2 + 24x - 3}{(2x+3)^2}$ 3. $\dfrac{42x}{(3x^2+1)^2}$ 4. $\dfrac{1}{(1-x)^{\frac{3}{2}}\sqrt{x+1}}$

5. $\dfrac{4(x^3-2)(x^3+1)}{x^3}$ 6. $\dfrac{-6(x^2+1)}{x^4\sqrt{3x^2+2}}$ 7. $\dfrac{-8x^2 + 12x + 24}{3(2x+1)^3(x^2-x-4)^{\frac{2}{3}}}$ 8. $\dfrac{10x^3 - 2x^2 - x + 4}{2(2-x)^{\frac{3}{2}}}$

9. $\dfrac{-6x}{(1-3x^2)^2}$ 10. $\dfrac{4}{(x-2)^2}$ or $\dfrac{4}{(2-x)^2}$

Exercise 6

1. $\dfrac{3t}{2}$ 2. $\dfrac{2(4-t^3)^2}{3t}$ 3. $\dfrac{2r}{3(l+2r)}$ 4. $\dfrac{2t-1}{10t-1}$ 5. $\dfrac{-2m(v-u)}{(u+v)t^2}$ 6. $\dfrac{5(t^4+1)^2}{3(t^2-1)^2}$

7. $\dfrac{1-12t^2}{2t}$ 8. $\dfrac{15t^2}{2}$ 9. $\dfrac{8}{r}$ 10. $\dfrac{2s^3}{3}$

Exercise 7

1. $\dfrac{-15x^2}{3-2y}$ or $\dfrac{15x^2}{2y-3}$ 2. $\dfrac{-4x}{9y^2}$ 3. $\dfrac{3y^2 + 12x^2 - y}{x - 6xy}$ 4. $\dfrac{3x^2 - y^2 - 4x}{2xy}$ 5. $\dfrac{x^2}{y}$

6. $\dfrac{4xy}{2x^2+5}$ 7. $\dfrac{-2xy}{x^2 + 6y^2 - 1}$ 8. $\dfrac{4y}{10y - 4x - 1}$ 9. $\dfrac{2xy^2 - 1}{2y(x^2-1)}$ 10. $\dfrac{-5x^4 y^2}{3}$

Exercise 8

1. $2\sec^2 2x$ 2. $-\dfrac{1}{5}\sin\dfrac{1}{5}x$ 3. $-50\sin 5x$ 4. $3\sin^2 x\cos x$ 5. $2\sec^2 x\tan x$

6. $60x^4\sin^3 3x^5\cos 3x^5$ 7. $-12x\cot 6x^2\csc 6x^2$ 8. $10x^4\sec 2x^5\tan 2x^5$ 9. $6x\sec^2 3x^2$

10. $-x(3x\sin 3x - 2\cos 3x)$ 11. $\dfrac{6x\cos 2x - 9\sin 2x}{x^4}$ 12. $48x^3\sec x^4\tan x^4$

13. $\dfrac{-5\tan x\cos 5x - \sec^2 x(2 - \sin 5x)}{\tan^2 x}$ 14. $\dfrac{\sin^2 x(3x\cos x - \sin x)}{2x^2}$

15. $\dfrac{-(2x+6)\csc^2 2x - \cot 2x)}{(x+3)^2}$ 16. $\dfrac{-\sin\sqrt[3]{3}}{5x^2}$ 17. $\dfrac{2[x - (x^2-3)\tan 2x]}{\sec 2x}$

18. $6x\cos 3x^2 - 18x^3\sin 3x^2$ 19. $-\cos x(2\cot x\sin x + \cos x\csc^2 x)$

20. $\dfrac{2(\sin^2 x + \sin^2 x)}{(\sin x + \cos x)^2}$ or $\dfrac{2}{(\sin x + \cos x)^2}$ 21. $\dfrac{4\sec 4x\sin(4x-2) + \tan 4x\cos(4x-2)}{\cos^2(4x-2)}$

22. $-3\sin 3x - x^2\sec x\tan x - 2x\sec x$ 23. $27x^2\cos x^3\sin^8 x^3$ 24. $\dfrac{3\cos 6\sqrt{x}}{\sqrt{x}}$

25. $3\cos x^3 - \dfrac{2\sin x^3}{x^3}$ 26. $-30x^4\cos^2 2x^5\sin 2x^2$ 27. $-10\sin 2x\sin 10x - 2\cos 2x\cos 10x$

28. $6x^2\cos x^4 - \dfrac{3\sin x^4}{2x^2}$ 29. $\dfrac{(10x - 5)\sec^2 5x - 2\tan 5x}{(2x-1)^2}$ 30. $\dfrac{\sec^2 x\tan x}{x} - \dfrac{\tan^2 x}{2x^2}$

Exercise 9

1. $\dfrac{1}{3\sqrt[3]{7y^2}}$ 2. $\dfrac{3y^2}{2\sqrt{y^3-1}}$ 3. $\dfrac{1}{5\sqrt[5]{2(y+3)^4}}$ 4. $\dfrac{1}{\sqrt[3]{2(3y+27)^2}}$ 5. $\dfrac{-1}{(y-2)^2}$

6. $\dfrac{-3}{2y^2\sqrt{\frac{3}{y}+5}}$ 7. $2y^3$ 8. $5y^4$ 9. $\dfrac{-5}{(y-4)^2}$ 10. $\dfrac{-1}{3y^2\sqrt[3]{(\frac{1}{y}+8)^2}}$

Exercise 10

1. $\dfrac{2}{\sqrt{1-4x^2}}$ 2. $\dfrac{-1}{\sqrt{1-x^2}}$ 3. $\dfrac{-6x}{9x^4+1}$ 4. $\dfrac{-4x^3}{x^8+1}$ 5. $3\sec^{-1}x+\dfrac{3}{\sqrt{x^2-1}}$

6. $2x+\dfrac{3}{x^2+1}$ 7. $\dfrac{1}{(x+5)^2+1}$ or $\dfrac{1}{x^2+10x+26}$ 8. $\dfrac{1}{x^2\sqrt{1-\frac{1}{x^2}}}$ 9. $\dfrac{-1}{5\sqrt{1-y^2}}$

10. $\dfrac{1}{2y\sqrt{y-1}}$ 11. $5\tan^{-1}3x+\dfrac{15x}{9x^2+1}$ 12. $2x+\dfrac{20x^4}{\sqrt{1-x^{10}}}$ 13. $\dfrac{-3}{x\sqrt{x^6-1}}$

14. $\dfrac{1}{2(y^2+1)}$ 15. $\dfrac{1}{15y^{\frac{2}{3}}\sqrt{1-y^{\frac{2}{3}}}}$

Exercise 11

1. $3\cosh 3x-2\sinh x$ 2. $-5x\,\text{sech}\,x(x\tanh x-2)$ 3. $\dfrac{2(x\cosh 2x-\sinh 2x)}{3x^3}$

4. $6\cosh 3x\sinh 3x$ 5. $60x^4\cosh^2 4x^5\sinh 4x^5$ 6. $-10x^4\text{cosech}\,2x^5$

7. $-12x\,\text{sech}^3 2x^2\tanh 2x^2$ 8. $x(5x\sinh 5x+2\cosh 5x)$

9. $\dfrac{3\text{cosech}^2 3x\tanh 5x+5\coth 3x\,\text{sech}^2 5x}{\coth^2 3x}$ 10. $\dfrac{2\cosh^2 x(3x\sinh x-\cosh x)}{3x^2}$

Exercise 12

1. $\dfrac{1}{x}\log_a e$ 2. $\dfrac{3x^2}{x^3+5}\log_a e$ 3. $\dfrac{36x^2}{2x^3-5}\log_a e$ 4. $\dfrac{2}{3x}\log_a e$ 5. $\dfrac{4x}{(x^2+1)(1-x^2)}\log_a e$

or $\dfrac{-4x}{(x^2+1)(x^2-1)}\log_a e$ or $\dfrac{-4x}{(x^4-1)}\log_a e$ 6. $\dfrac{15x^2}{5x^3-1}\log_5 e$ 7. $\dfrac{-2}{x}\log_2 e$ 8. $\dfrac{2x}{x^2+3}$

9. $\dfrac{3\ln^2 x}{x}$ 10. $\dfrac{5x^4}{2x^5-1}$ 11. $2x^3(4\ln x+1)$ 12. $\dfrac{-28}{3-7x}$ or $\dfrac{28}{7x-3}$

13. $6x\ln(4x^3+1)+\dfrac{36x^4}{4x^3+1}$ 14. $\dfrac{2-4\ln x}{x^3}$ 15. $\dfrac{-30x}{1-5x^2}$ or $\dfrac{30x}{5x^2-1}$

Exercise 13

1. $10a^{2x}\log_e a$ 2. $a^{5x^2-x}\ln a(10x-1)$ 3. $a^{3x}x^3(3x\ln a+4)$

4. $-2e^{-2x}-3e^{-x}$ or $-e^{-2x}(2+3e^x)$ 5. $\dfrac{\sqrt{3}\,e^{\sqrt{3x}}}{\sqrt{x}}$ 6. $6x^2e^{x^2}(2x^2+3)$

7. $\dfrac{5\ln(a^{10x}-1)}{a^{5x}}$ 8. $\dfrac{6e^{4x}(2x\ln 2x^2+1)}{x}$ 9. $\dfrac{3x\sqrt{e^{5x}}(5x+4)}{2a}$ 10. $x^3(\log_{10}e+4\log_{10}7x)$

11. $3\ln6(6^{3x})$ 12. $2x - 3(e^{x^2 - 3x})$ 13. $\dfrac{-(5x + 3)e^{\frac{3}{x}}}{5x^7}$ 14. $6e^{-3x}$

15. $3(2x + 1)(x^2 + x)^2 e^{(x^2 + x)^3}$ 16. $2^{x^2} 2x\ln2$ 17. $2a^{2x}\ln a$ 18. $\dfrac{e^x(x\ln10x^3 + 3)}{x}$

19. $\dfrac{\sqrt{5}(4x + 1)e^{2x}}{2\sqrt{x}}$ 20. $2x^4(4\log_{10}e + 5\log_{10}5x^4)$

Exercise 14

1. $x^{2x}(2 + 2\ln x)$ 2. $\dfrac{3x^{\ln 2x}(\ln 2x + \ln x)}{x}$ 3. $\ln(1 - 4x^2) - \dfrac{8x^2}{1 - 4x^2}$

4. $(6x^x - 5)^{3x}\left[3\ln(6x^2 - 5) + \dfrac{36x^2}{6x^2 - 5}\right]$ 5. $\dfrac{(2x^3 + 1)(x - 2)^3}{3x^2(x^3 - 1)^2}\left(\dfrac{6x^2}{2x^3 + 1} + \dfrac{3}{x - 2} - \dfrac{2}{x} - \dfrac{6x^2}{x^3 - 1}\right)$

6. $\dfrac{(2x - 1)(x^2 - 2)}{(1 - x)(x - 3)^2}\left(\dfrac{2}{2x - 1} + \dfrac{2x}{x^2 - 2} + \dfrac{1}{1 - x} - \dfrac{2}{x - 3}\right)$

7. $15x^3 x^{3x^4}(4\ln x + 1)$ or $15x^{3x^4 + 3}(4\ln x + 1)$ 8. $3e^{3x}(e^{e^{3x}})$ or $3e^{e^{3x} + 3x}$

9. $2^{e^x}e^x\ln2$ 10. $\dfrac{x^{\ln(x^2 - 4x)}[(x - 4)\ln(x^2 - 4x) + (2x - 4)\ln x]}{x(x - 4)}$ 11. $\ln(2x^2 + 10) + \dfrac{4x^2}{2x^2 + 10}$

12. $(3x^2 - 8)^{2x}\left[2\ln(3x^3 - 8) + \dfrac{18x^3}{3x^3 - 8}\right]$ 13. $\dfrac{(x + 3)^2(x - 3)^2}{x^3 - 1}\left(\dfrac{2}{x + 3} + \dfrac{2}{x - 3} - \dfrac{3x^2}{x^3 - 1}\right)$

14. $15x^{(\ln x)^2}(\ln x)^2$ 15. $30x^{3x^2}x(2\ln x + 1)$ or $30x^{3x^2 + 1}(2\ln x + 1)$

Exercise 15

1. $\dfrac{3}{5x^2}$ 2. $\dfrac{5xe^{5x}}{2}$ 3. $\dfrac{-2x\sin x^2}{\cos x}$ 4. $\dfrac{x(6x - 5)}{x - 1}$ 5. $\dfrac{14x^2}{3}$ 6. $\dfrac{2a^x\ln a}{e^x}$

7. $\dfrac{-2\tan x \sec^2 x}{5\sin 5x}$ 8. $\dfrac{1}{(x - 1)(e^{x - 1})}$ 9. $\dfrac{1}{xe^{5x}}$ 10. $\dfrac{e^x + e^{-x}}{(e^x - e^{-x})(1 + \ln x)}$

Exercise 16

1. $\dfrac{dy}{dx} = 2x(2x^2 - 3x - 7)$, $\dfrac{d^2y}{dx^2} = 2(6x^2 - 6x - 7)$, $\dfrac{d^3y}{dx^3} = 2(12x - 6)$ 2. $-\dfrac{3}{x^2}$

3. $20x^2 e^{5x^4}(20x^4 + 3)$ 4. $-50x^8\sin^2 x^5 + 40x^3\cos x^5\sin x^5 + 50x^8\cos x^5$ 5. $\dfrac{128}{(4x - 10)^2}$

6. $\dfrac{-(x^2 - 2)\sin x - 2x\cos x}{x^3}$ 7. $\dfrac{4\ln x - 6}{x^3}$ 8. $-2\cosec^2 x(\cosec^2 x + 2\cot^2 x)$

9. $2\cos^2 x - 3\sin^2 x - 5\cos 2x$ 10. 0

11. $-8x^2\sin x^2 + 4\cos x^2 + 2\sin^2 x - 2\cos^2 x - 4x\cos x^2 + 2\cos x\sin x$

12. $\dfrac{2[(x^2\sec^2 x + 1)\tan x - x\sec^2 x]}{3x^3}$ 13. $e^{-x}(4\sin 2x - 3\cos 2x)$ 14. 0

15. $\dfrac{-(9x^2 - 2)\sin 3x - 6x\cos 3x}{3x^3}$

Exercise 17

1. $2x(3x^4 - 5)^3 + 36x(3x^4 - 5)^3$ or $2x(3x^4 - 5)^2(21x^4 - 5)$ 2. 209

3(a) $\dfrac{\sqrt{3x^2 - 1}\,(3x^2 + 2)}{x^3}$ (b) $5\sqrt{2}$ 4(a) $\dfrac{12x^2 - 3y^3}{9xy^2 - 2y - 5}$ (b) $-\dfrac{9}{16}$ 5. $-\dfrac{2}{y}$

6. $3x(x^2 + 5)^2(3x^3 + 5x - 2)$ 7. $e^{\sin x + \tan x}(\sec^2 x + \cos x)$ 8. $\dfrac{2x(\sin x - x\cos x)}{\sin^3 x}$

9. $\dfrac{2\cos x \sin x - 3\sin 3x}{\sin^2 x + \cos 3x}$ 10. $-5a^{\cos 5x}\sin 5x \ln a$ 11. $5x^2 e^{-3x} - 10xe^{-3x} + 2e^{-3x}$

12. $-(2x^3 + 2)\cos\left(\dfrac{1 - 2x^3}{x^2}\right)$ 13. $\dfrac{4x^4 - 6x^3 + 12x^2 - 18x + 6}{x^3}$ 14. 0

15. $5\left(\dfrac{2}{x^2} - \dfrac{2}{x^3}\right)\left(\dfrac{1}{x^2} - \dfrac{2}{x}\right)^4$ or $\dfrac{10(x-1)(1-2x)^4}{x^{11}}$ 16(a) $\dfrac{3x^2 e^{x^3 - y^3} - y}{3y^2 e^{x^3 - y^3} + x}$ (b) 1

17. 0   18. $\dfrac{2\sin\frac{1}{x^2}}{x^2} - \dfrac{4\cos\frac{1}{x^2}}{x^4} + 2\cos\dfrac{1}{x^2}$ 19(a) $\dfrac{2e^x(e^x + 1)}{(e^x - 1)^3}$

(b) $\dfrac{4(e^{2x} - e^{-x}) + (1 + e^{-x})(e^x - 1)^3}{(e^x - 1)^3(1 - e^{-x})}$ 20. $2\tan x$ 21. 0

22. $3\left(\dfrac{\sqrt{5}}{4\sqrt{x}} + 3\right)\left(3x + \dfrac{\sqrt{5x}}{2}\right)^2$ 23. $\dfrac{-10x - 18}{x^4}$ or $\dfrac{-2(5x + 9)}{x^4}$

24. $9x^2(x^3 + 1)^2 + \dfrac{5 + 4x - 10x^2}{(x^2 + 1)^2}$ 25. 0

Exercise 18

1.

Mark	65	66	67	68	69	70	71	72	73	74
Frequency	4	2	6	5	6	3	3	2	3	5

2.

Score	5	6	7	8	9
Frequency	5	7	2	2	4

3.

No of Seed	20	21	22	23	24	25	26	27	28	29	30
Frequency/No of Pod	6	4	3	5	2	2	1	4	7	3	3

4.

Age	10	11	12	13	14	15
No of School	5	4	5	5	6	5

Exercise 19

1. Mean = 4   Median = 3   Mode = 2
2. (a) 24   (b) 25   (c) 25
3. Mean = 111.2   Median = 111.5   Mode = 105
4. Mean = 4.3   Median = 4.5   Mode = 5
5. (a) 5.1   (b) 5   (c) 4
6. (a) 13.2   (b) 13.5   (c) 14
7. (a) 11   (b) 58kg

Exercise 20

1.

Score	40 − 44	45 − 49	50 - 54	55 - 59	60 − 64
Frequency	8	14	13	9	6

2.

Weight	21 - 25	26 - 30	31 - 35	36 - 40	41 - 45	46 - 50	51 − 55
Frequency	2	9	8	2	8	2	9

3.

Class interval	Frequency	Class Boundary	Class width	Class mid-value
5 − 9	2	4.5 − 9.5	20	7
10 − 14	5	9.5 − 14.5	20	12
15 − 19	5	14.5 − 19.5	20	17
20 − 24	7	19.5 − 24.5	20	22
25 − 29	1	24.5 − 29.5	20	27

4.

Class interval	Frequency	Class Boundary	Class width	Class mid-value
1 – 20	1	0.5 – 20.5	20	10.5
21 – 40	4	20.5 – 40.5	20	30.5
41 – 60	7	40.5 – 60.5	20	50.5
61 – 80	3	60.5 – 80.5	20	70.5
81 – 100	5	80.5 – 100.5	20	90.5

5.

Class interval	Frequency	Class Boundary	Class width	Class mid-value
0 – 90	2	-5 – 95	100	45
100 – 190	4	95 – 195	100	145
200 – 290	1	195 – 295	100	245
300 – 390	7	295 – 395	100	345
400 – 490	1	395 - 495	100	445

Exercise 21

1. (a) 72.5　　(b) 70.9　　(c) 67.9

2. (a) 90.4　　(b) 91.5　　(c) 92.7

3. (a) 44.1　　(b) 41.7　　(c) 36.9

Exercise 22

1. 1.67　　　2. 1.82　　　3. 1.6　　　4. 16.27　　　5. 12.5　　　6. 9.1　　　7. 2.7

8. (a) The frequency table is as shown below.

No on Die	1	2	3	4	5	6
No of Times	7	7	9	11	7	9

(b) 1.41

Exercise 23

1. Variance = 2　　　Standard Deviation = 1.41

2. Variance = 6　　　Standard Deviation = 2.45

3. (a) 199.04　　　(b) 14.11

4. (a) 40.69　　　(b) 6.38

5. (a) 1.049　　　(b) 1.024

Exercise 24

1. The frequency table is as shown below.

Mark	11 - 20	21 – 30	31 - 40	41 – 50	51 - 60	61 - 70	71 - 80	81 - 90	91-100
No of Student	4	5	4	5	8	3	5	3	3

(a) 53   (b) 33   (c) 72.5   (d) 39.5   (e) 19.75   (f) 38   (g) 64.5

2. (a) 32.36   (b) 25.18   (c) 40.25   (d) 15.07   (e) 7.54   (f) 41.6

(g) 37.33   (h) 22.06

3. (a) 1.7   (b) 1.23   (c) 2.31   (d) 1.60   (e) 0.98

4. (a) 17.5   (b) 11.29   (c) 22.56   (d) 11.27   (e) 5.63   (f) 20.06   (g) 29.5

5. (a) 1.54   (b) 0.83   (c) 2.08   (d) 0.83   (e) 1.01

Exercise 25   There is no exercise in chapter 25.

Exercise 26

1. (a) $\frac{1}{10}$ (b) $\frac{9}{20}$ (c) $\frac{1}{4}$ (d) $\frac{3}{4}$ (e) $\frac{7}{20}$

2. $\frac{1}{4}$ 3. $\frac{4}{13}$ 4. $\frac{2}{5}$ 5. $\frac{4}{5}$

6. (a) $\frac{3}{5}$ (b) $\frac{2}{5}$

7. $\frac{11}{25}$ 8. $\frac{7}{10}$

9. (a) $\frac{2}{5}$ (b) $\frac{1}{2}$ (c) $\frac{9}{10}$ (d) $\frac{3}{10}$ (e) $\frac{9}{20}$

10. (a) $\frac{1}{26}$ (b) $\frac{3}{26}$ (c) $\frac{2}{13}$ (d) $\frac{4}{13}$ (e) $\frac{4}{13}$

11. $\frac{2}{11}$

Exercise 27

1. $\frac{1}{13}$ 2. $\frac{1}{26}$ 3. $\frac{1}{26}$ 4. 1   5. $\frac{1}{26}$ 6. $\frac{12}{13}$

7. (a) $\frac{2}{13}$ (b) $\frac{2}{13}$ (c) $\frac{5}{52}$ (d) $\frac{7}{26}$ (e) $\frac{15}{52}$ (f) $\frac{7}{13}$

8. (a) $\frac{1}{169}$ (b) $\frac{1}{13}$ (c) $\frac{6}{169}$ (d) $\frac{1}{4}$ (e) $\frac{1}{13}$ (f) $\frac{12}{13}$

9. (a) $\frac{8}{663}$ (b) $\frac{8}{663}$ (c) $\frac{1}{221}$ (d) $\frac{1}{17}$ (e) $\frac{25}{102}$ (f) $\frac{13}{102}$ (g) $\frac{1}{221}$

10. (a) $\frac{37}{2197}$ (b) $\frac{2196}{2197}$

11. (a) $\frac{73}{5525}$ (b) $\frac{5452}{5525}$

Exercise 28

1. (a) $\frac{1}{2}$ (b) $\frac{1}{2}$ (c) 1

2. (a) $\frac{1}{4}$ (b) $\frac{3}{4}$ (c) $\frac{1}{4}$ (d) $\frac{3}{4}$ (e) $\frac{1}{4}$

3. (a) $\frac{1}{8}$ (b) $\frac{7}{8}$ (c) $\frac{1}{8}$ (d) $\frac{7}{8}$ (e) $\frac{1}{2}$ (f) $\frac{7}{8}$

4. (a) $\frac{15}{16}$ (b) $\frac{1}{16}$ (c) $\frac{11}{16}$ (d) $\frac{15}{16}$ (e) $\frac{3}{8}$

5. $\frac{31}{32}$

Exercise 29

1. (a) $\frac{1}{6}$ (b) $\frac{1}{6}$ (c) $\frac{1}{6}$ (d) $\frac{1}{2}$ (e) $\frac{5}{6}$ (f) $\frac{5}{6}$

2. (a) $\frac{1}{2}$ (b) $\frac{1}{2}$ (c) $\frac{5}{6}$ (d) $\frac{1}{2}$ (e) $\frac{2}{3}$ (f) $\frac{1}{3}$

3. (a) $\frac{1}{12}$ (b) $\frac{1}{12}$

4. (a) $\frac{11}{36}$ (b) $\frac{13}{18}$ (c) $\frac{11}{12}$ (d) $\frac{7}{12}$ (e) $\frac{1}{36}$ (f) $\frac{7}{18}$ (g) $\frac{7}{18}$

5. (a) $\frac{1}{6}$ (b) $\frac{5}{12}$ (c) $\frac{1}{36}$ (d) $\frac{1}{6}$ (e) $\frac{1}{6}$

6. $\frac{1}{6}$

Exercise 30

1. (a) $\frac{1}{4}$ (b) $\frac{2}{5}$ (c) $\frac{7}{20}$ (d) 0   (e) $\frac{13}{20}$ (f) $\frac{13}{20}$

2. (a) $\frac{7}{8}$ (b) $\frac{3}{4}$ (c) $\frac{1}{4}$

3. (a) $\frac{2}{125}$ (b) $\frac{8}{125}$ (c) $\frac{92}{125}$ (d) $\frac{31}{125}$

4. (a) $\frac{27}{64}$ (b) $\frac{37}{64}$ (c) $\frac{63}{64}$ (d) $\frac{27}{64}$

5. (a) $\frac{3}{8}$ (b) $\frac{1}{2}$ (c) $\frac{3}{8}$ (d) $\frac{1}{2}$ (e) $\frac{1}{2}$

6. (a) $\frac{15}{364}$ (b) $\frac{145}{182}$ (c) $\frac{25}{91}$

7. (a) $\frac{2}{5}$ (b) $\frac{8}{15}$

8. (a) $\frac{7}{22}$ (b) $\frac{7}{11}$ (c) $\frac{21}{22}$

Please if you found this book well simplified enough for easier understanding, kindly give it a five star rating on amazon so as to encourage people to buy this book, thereby putting more money in my pocket and helping students improve on their skills on algebra and differential calculus. Thank you.

If you want to see other books written by the author, just simply search for the author's name, Kingsley Augustine on amazon.com

If you have any enquiry, suggestion or information concerning this book, please contact the author through the email below.

KINGSLEY AUGUSTINE

kingzohb2@yahoo.com